Key issues in agricultural ethics

It is widely recognised that agriculture is a significant contributor to global warming and climate change. Agriculture needs to reduce its environmental impact and adapt to current climate change whilst still feeding a growing population, i.e. become more 'climate-smart'. Burleigh Dodds Science Publishing is playing its part in achieving this by bringing together key research on making the production of the world's most important crops and livestock products more sustainable.

Based on extensive research, our publications specifically target the challenge of climate-smart agriculture. In this way we are using 'smart publishing' to help achieve climate-smart agriculture.

Burleigh Dodds Science Publishing is an independent and innovative publisher delivering high quality customer-focused agricultural science content in both print and online formats for the academic and research communities. Our aim is to build a foundation of knowledge on which researchers can build to meet the challenge of climate-smart agriculture.

For more information about Burleigh Dodds Science Publishing simply call us on +44 (0) 1223 839365, email info@bdspublishing.com or alternatively please visit our website at www.bdspublishing.com.

Related titles:

Assessing the environmental impact of agriculture
Print (ISBN 978-1-78676-228-3); Online (ISBN 978-1-78676-230-6, 978-1-78676-231-3)

The sustainable intensification of smallholder farming systems
Print (ISBN 978-1-78676-430-0); Online (ISBN 978-1-78676-432-4, 978-1-78676-433-1)

Genome editing for precision crop breeding
Print (ISBN 978-1-78676-447-8); Online (ISBN 978-1-78676-449-2, 978-1-78676-450-8)

Understanding the behaviour and improving the welfare of chickens
Print (ISBN 978-1-78676-422-5); Online (ISBN 978-1-78676-424-9, 978-1-78676-425-6)

Understanding the behaviour and improving the welfare of pigs
Print (ISBN 978-1-78676-443-0); Online (ISBN 978-1-78676-444-7, 978-1-78676-445-4)

Understanding the behaviour and improving the welfare of dairy cattle
Print (ISBN 978-1-78676-459-1); Online (ISBN 978-1-78676-461-4, 978-1-78676-462-1)

Developing smart agri-food supply chains: Using technology to improve safety and quality
Print (ISBN 978-1-78676-749-3); Online (ISBN 978-1-78676-751-6, 978-1-78676-752-3)

Chapters are available individually from our online bookshop:
https://shop.bdspublishing.com

climate•SMART•publishing
IN AGRICULTURAL SCIENCE

BURLEIGH DODDS SERIES IN AGRICULTURAL SCIENCE
NUMBER 140

Key issues in agricultural ethics

Edited by Professor Emeritus Robert L. Zimdahl, Colorado State University, USA

burleigh dodds
SCIENCE PUBLISHING

Published by Burleigh Dodds Science Publishing Limited
82 High Street, Sawston, Cambridge CB22 3HJ, UK
www.bdspublishing.com

Burleigh Dodds Science Publishing, 1518 Walnut Street, Suite 900, Philadelphia, PA 19102-3406, USA

First published 2023 by Burleigh Dodds Science Publishing Limited

Library of Congress Control Number: 2023938359

British Library Cataloguing in Publication Data
A catalogue record for this book is available from the British Library

ISBN 978-1-80146-313-3 (print)
ISBN 978-1-80146-315-7 (PDF)
ISBN 978-1-80146-314-0 (ePub)
ISSN 2059-6936 (print)
ISSN 2059-6944 (online)

DOI: 10.19103/AS.2023.0125

Typeset by Deanta Global Publishing Services, Dublin, Ireland

Contents

Part 2 Ethical issues

Series list

Introduction

Agriculture is facing unprecedented scrutiny for its social and environmental effects. Many of the key choices it must make are fundamentally about ethics. This volume explores key ethical debates surrounding agriculture and agri-food supply chains.

The chapters are divided in two parts: Part 1 provides a background on what agricultural ethics is, whilst also identifying current approaches to the field of ethics. Chapters also review how agricultural ethics has not been institutionalised in US land grant universities, gender dimensions of agricultural ethics, the farmer's perspective of agricultural ethics and the major aspects of the virtue ethics tradition, specifically drawing attention to Wendell Berry. Chapters in Part 2 give specific examples of ethical issues in agriculture, such as the environment, migrant labour and sustainable intensification. Chapters on key ethical issues in livestock farming, developing new technologies, intellectual property rights and ethics of agricultural research are also included.

Part 1 General

The book opens with a chapter that describes what agricultural ethics is and why it matters. Chapter 1 begins by first providing an overview of the key concepts of ethics and common morality, then moves on to review common morality in agriculture and food systems. The chapter also highlights some relevant topics in agricultural ethics, before reviewing the various social science research methods currently involved in analysing relevant agricultural ethics data. A section on why agricultural ethics matters is also included, focusing on how the outcome of agricultural ethics debates affect the work of those involved and why it is important for these people to strive to do what is ethical. A discussion on whether agricultural ethics can be taught is also provided.

The next chapter of Part 1 focuses on current approaches to ethics. Chapter 2 first describes the field of ethics, then moves on to analyse the current three main schools of thought in contemporary ethics through their own individual sections: utilitarianism, deontology and virtue ethics. For each school of thought, the chapter focuses on both the anthropocentric and nonanthropocentric theories involved and highlights their differences. The chapter then moves on to debate if whether these three schools of thought should be considered as competing theories or as part of an ethical toolkit where they can be used together.

Chapter 3 draws attention to lack of institutionalisation of agricultural ethics in US land grant universities. The chapter begins by providing an overview of food production and 'productivism', the ethical position that characterises agriculture. It then moves on to review the ethical standards among people in agriculture and the current scientific challenges that agriculture faces. The chapter also highlights current problems facing agricultural systems, agriculture's ethical dilemmas and provides an ethical perspective on agriculture.

The next chapter provides an important overview of equity focused research aimed at better understanding women's contributions in food production. Chapter 4 specifically presents a detailed overview of gender-sensitive research in agriculture, since its initial development in the 1970s and 1980s. While women contribute substantially to agricultural production, there are unique challenges and barriers that affect women. The chapter includes an overview of research performed in gender and development circles, as well as work on the gender dimensions of agricultural in the global North. It ends by discussing emerging themes in the literature.

Chapter 5 looks at agricultural ethics from the farmer's perspective. The chapter first highlights the importance of understanding farmers' constraints and motivations as well as their responsibilities within the agri-food supply system. It moves on to review the environmental and social costs that occur as a result of agricultural production practices, which is followed by an overview of current agricultural issues faced by farmers. The chapter also addresses agriculture's image within society and how this can be changed, as well as the public fear associated with the effect of agriculture on the environment. Sections on the importance of conservation and sustainability are also provided.

The final chapter of Part 1 focuses on virtue ethics, Wendell Berry and agriculture. Chapter 6 begins by detailing the major aspects of the virtue ethics tradition, particularly as they are understood by Aristotle. The chapter then turns to Wendell Berry's thought and argue that Berry embodies well many of the facets of this tradition. It considers three virtues in Berry's thought: humility, generosity, and love. The chapter reveals by taking these virtues seriously, Berry offers the agricultural community a more sustainable path forward. Berry's advocacy of virtues in agriculture offers wisdom to researchers and practitioners alike which, the chapter argues, can ameliorate many present ecological and economic concerns.

Part 2 Ethical issues

The first chapter of Part 2 focuses on the ethical issues of agriculture and the environment. Chapter 7 begins by providing an overview of what has happened to the environment as a consequence of agricultural practices, then moves on to review how these practices have changed over the years. The chapter also reviews the constraints to ethical awareness and moral action, the ability to make judgements in terms of agricultural practices and how its concepts can be applied in practice. A section on environmental ethics and how this could potentially help to improve the situation is also included, followed by an overview of ethics and worldview transformation. Worldviews, experiential learning and systemic development are also reviewed. The chapter also provides a discussion on the Australian aboriginal perspective of ethics, agriculture and the environment.

Chapter 8 examines the ethical issues and challenges of migrant labour in agriculture. The chapter first describes the importance of migrant workers in the agricultural and agri-food sectors, whilst also highlighting the policies regulating their role and conditions of employment. The chapter particularly focuses on migrant workers within the European Union (EU), using the example of Italy as a basis for discussion to illustrate issues and policies at a national level. It then moves on to review the ethical challenges in forming fair and sustainable polices to support migrant workers in agriculture and the wider agri-food supply chain.

The subject of Chapter 9 is the ethical issues of sustainable intensification in agriculture. The chapter begins by highlighting the importance of sustainable intensification in agriculture, then moves on to examine four policy areas, their goals as well as ethical trade-offs in relation to the use of sustainable intensification. The chapter focuses specifically on biodiversity versus competing land use, protein intensification versus human–animal relations, quantity versus quality of food and using land efficiently versus vibrant rural communities. A section on the pragmatic responses to intensification-related tensions is provided.

Chapter 10 focuses on the key ethical issues in livestock farming. The chapter addresses the practical application of animal ethics in agriculture as necessary for animal ethics and for those who work with animals in agriculture. The chapter also provides an overview of the main positions in theoretical animal ethics, then moves to frame a discussion of the continuum of positions on the moral acceptability of the use of non-human animals for human ends. It also outlines ethical issues arising from the use of animals in agriculture, organised into three groups: issues directly related to non-human animals, issues related to humans and society and issues related to the natural environment.

The final chapter of the book reviews ethical issues of developing new technologies in agriculture. Chapter 11 first examines the role of ethics in the development of radical new technologies in general. The chapter then analyses the Rights and Resources Initiative and how it can be applied in agriculture. Four potential ethical issues are also identified for developing new technologies in agriculture. The chapter also provides two examples of developing agricultural technologies that aim to change organisms, the first being genetic modification and the second the toolbox of New Plant Breeding Technologies.

Acknowledgements

We wish to acknowledge the following for their help in reviewing particular chapters:

- Chapter 3: Dr Donald Viney, Professor of Philosophy, Pittsburg State University, USA, Dr Wade Casey, Loyola University Chicago, USA and Dr Thomas Holtzer, Colorado State University, USA
- Chapter 5: Fargo Chamber of Commerce Distinguished Professor Emeritus David Danbom, formerly North Dakota State University, USA

Part 1

General

Chapter 1

What is agricultural ethics and why does it matter?

Paul B. Thompson, Michigan State University, USA*

1 Introduction

The title of this chapter illustrates the curious grammar of the word *ethics*. Native speakers of English commonly speak of ethics as something that people have or lack. In this sense, ethics are rules of behavior or patterns of conduct. Readers should notice the shift from the plural 'are' in the previous sentence to the verb 'is' in the title of the chapter. The meaning of the word 'ethics' changes from naming a collection of related practices, activities or principles to indicating a singular practice or activity. While this might seem like a trivial observation, it is indicative of a deeper confusion that has dogged the reception of Robert Zimdahl's important work on agricultural ethics. This explains why there is a need for a chapter called 'What is agricultural ethics?'

This chapter will work through some meanings of the word 'ethics' as it might be applied in the context of agriculture and food systems. Readers will arrive at my answer to the question 'What is agricultural ethics?' at the endpoint

* Paul B. Thompson, Professor of Philosophy, of Community Sustainability and of Agricultural Food and Resource Economics and W.K. Kellogg Professor Of Agricultural, Food and Community Ethics, Department of Philosophy, Michigan State University, USA

http://dx.doi.org/10.19103/AS.2023.0125.01

of this journey, but here is a start: Agricultural ethics is a specific discipline for inquiry into the myriad normative issues that interpenetrate every aspect of agricultural production and food systems. As understood in the specific sense developed here, agricultural ethics matters because some unique features in the institutionalization of agricultural science and education have created gaps in the thought processes that support policy and the innovation process for agricultural technologies. The chapter concludes with some remarks on teaching agricultural ethics.

2 Key concepts: ethics and common morality

The twentieth-century philosopher R. M. Hare (1919–2002) recognized a distinction between what he called *common morality* and philosophical ethics. Common morality, Hare said, is what your grandmother knew to be ethically correct. One does not need a philosopher to explain this. Everyone, Hare wrote, internalizes a set of rules or norms for regulating their own behavior and correlative expectations for the conduct of others. These norms are communicated and socially reinforced through words such as 'good', 'bad', 'right' and 'wrong' and through legal and religious institutions that codify and promote shared practice revolving around concepts of justice, virtue, honesty and loyalty. It is surprisingly difficult to account for the sources of these norms, but we should not doubt their existence or their binding authority over many domains of personal and social conduct. Similarly, the generally small cultural variations in a community's normative expectations should not obscure the core of shared norms in virtually every known human society. That is common morality (Hare, 1981).

Hare was hardly alone among philosophers and social theorists in holding such a view, but his understanding of the relationship between the cultural form of common morality and the activity of moral philosophers is helpful in the present context. As noted, most elements of common morality are stable across different cultural groups. These include norms of truth-telling and prohibitions against stealing, robbery and physical violence against others, except in well-specified contexts. However, there are both differences across cultures and change over time within any given society. As an example, slavery was seen as morally justified for a considerable period in Western history. By the time chattel slavery was the dominant source of labor in the agriculture of the American South, it was viewed as a dubious and regrettable but necessary practice. The perspective shifted finally to the judgment that slavery is morally unacceptable. This transition marks a change in the common morality that was of particular importance for agriculture. Hare argued that philosophical ethics plays a special role in these transformations. It is an activity that assembles and curates intellectual resources for questioning common morality on specific points. A philosopher or social critic deploys these resources to offer rationally based assessments of how or whether changes in common morality are warranted.

Many contemporary philosophers would accept the notion that there is little point in questioning those aspects of common morality that seem to raise no disagreement, but they would also agree that philosophical ethics can be helpful both in challenging problematic aspects of common morality and in resolving that challenge in a manner that is ethically progressive. Hare used complex reasoning to support his judgment that there is one best way to pursue questions at the philosophical level (a form of utilitarianism). There is continuing debate among philosophers on this matter. Some agree that a well-specified ethical theory should rule over analysis at the philosophical level, whether or not they agree with the specifics of Hare's approach. Others, including myself, think methods in philosophical ethics cannot be specified much beyond a general commitment to the exchange of reasons and a good faith effort to understand and then accept, modify or rebut the views of those with whom you, at first blush, disagree. (A more detailed discussion of methods follows later in the chapter.) In either view, progressive approximation of the morally correct response is possible, and fallibility is a pervasive feature of the human condition.

Other philosophical views create barriers to the critique and evaluation of common morality in agricultural universities and research institutes. The influential economist Glenn L. Johnson (1918–2003) thought many agricultural scientists had adopted a flawed positivist philosophy of science. In this view, normative principles have no place in science, so agricultural scientists simply refuse to discuss ethical norms bearing on their science or the practice of agriculture (Johnson, 1976). The view is flawed because the statement 'normative principles have no place in science' *is* a normative principle. Scientists cannot follow it without doing what the principle itself says that they should not do (Thompson, 2004). Alternatively, the training of applied scientists (and I would add, especially social scientists) encourages them to adopt a stance of critical disengagement from disagreements about the ethics of farming practice. There may be good reasons for adopting this stance, but the model proposed by Hare shows that an ability to withhold the expression of one's moral commitments in certain instances does not imply that moral inquiry is impossible. Like any inquiry, moral inquiry can be frustrating and end in failure, but one would hope the practice of science would equip practicing scientists to deal with the possibility of this kind of disappointment.

By this point, readers may have guessed where this is all leading. On the one hand, people who work in agriculture and its supporting industries are subject to the expectations of common morality. They can be said to fail ethically when they do not meet these expectations. On the other hand, agricultural ethics is a form of philosophical ethics. It is the deployment of the analytic, argumentative and discursive tools of philosophical ethics to questions arising in agricultural practices and in the science and policy domains that support the production, processing and distribution of food and fiber. As such, it differs from the ways in which common morality functions to regulate the conduct of farmers, researchers, public officials and others who fill various roles in the food system. Those of us who aim to practice the form of inquiry that constitutes agricultural ethics do not mean to suggest that our work is, in any respect, a replacement for common morality. Much of the time, talk of ethics in agriculture is going to advert back to common morality. No one needs a philosopher to come in and endorse the proscription of lying or theft. However, before considering why agricultural ethics matters, it will prove helpful to provide some examples of how common morality functions in agricultural situations.

3 Common morality in agriculture and food systems

Farmers, ranchers and people who work for input firms or companies that process and distribute agricultural commodities are expected to act in an ethical manner. They are not supposed to lie or steal, and they are supposed to engage in fair, respectful dealing when interacting with others. This is not exceptional; it is what is expected of everyone. However, people being what they are, such expectations are not universally fulfilled. Just as nearly everyone breaks the law by exceeding the speed limit now and then, almost everyone who works in food and agriculture probably commits small ethical failures now and then. Other failures are egregious and have serious consequences. Both types of failure lead people to call for renewed attention to ethics in food and agriculture, but it is doubtful that even egregious failures call for a philosophical form of agricultural ethics.

Violations of the common morality are revealed by journalists covering food and agriculture, by criminal investigators and by social scientists conducting studies of food and fiber production or other aspects of the global food system. For example, in 2014 Richard Marosi, a staff writer for *The Los Angeles Times* reported the abuse of field workers in the Mexican tomato export industry. The abuses included dangerous and unhealthy working and housing conditions, illegal withholding of earned wages and threats of violence (Marosi, 2014). In 2009, just 2 years after they had executed their chief food safety official on charges of corruption, the Chinese government revealed a conspiracy to contaminate supplies of infant formula. Thousands were sickened, and there were at least three deaths (Sharma and Paradkar, 2010). In 2020, a Scottish sheep farmer was convicted of contaminating jars of baby food as what he claimed to be a protest against unfair treatment by the food industry (Davies, 2020).

Listing examples of this sort could continue indefinitely. There are two points to note. First, these cases are not ethically complicated. It is easy to see that unethical conduct occurred. Second, such cases precipitate a call for renewed attention to ethics. Such calls imply a decline in compliance with the norms of common morality or a measurable increase in the frequency or severity of violations. If true, such trends can have complex underlying causes requiring a multidisciplinary analysis and measured social experimentation with remedial measures. There may or may not be value in having someone regarded as a moral authority make pronouncements about the unethical nature of such conduct, and there may or may not be a reason to introduce training sessions and classwork for practitioners under the heading of ethics. One can imagine research projects and compliance tactics being introduced under the rubric of agricultural ethics. However, the specifically normative or ethical dimension of such phenomena is not in question. We are talking about ethics in the plural as something that people have or lack, not ethics in the singular as a form of normative inquiry.

The upshot is that the people who call for attention to ethics may not be calling for ethics in the philosophical sense Hare and many philosophers understand. A related phenomenon unfolded in bioethics. The term *bioethics* was popularized by Van Rensselaer Potter (1911-2001) in the 1960s. He saw bioethics as a form of interdisciplinary inquiry intended to address normative uncertainties posed by population growth, environmental decline and new reproductive technologies (ten Have, 2012). Following the creation of institutional review boards (IRBs) intended to curtail researchers' abuse of human and animal subjects, Potter's vision was sublimated by researchers who interpreted bioethics as research ethics and understood research ethics solely in terms of the compliance measures introduced by IRBs (Rollin, 2006). This is not to say IRBs were unnecessary or imply they do not address ethical issues. Nevertheless, the singular focus on the enforcement of ethically uncontroversial norms has obscured the role Potter expected a critically reflective inquiry into the normative dimensions of applied biological systems to play.

4 Topics in agricultural ethics

As departments of bioethics became institutionalized in medical schools, hospitals and biomedical research institutes, the term has become implicitly understood as oriented to medicine and public health. In fact, some of the questions Potter associated with bioethics are questions in agricultural ethics. For Potter, the tension between global population growth and resource depletion was central. He was clear in stating that decisions about converting land to agricultural use and intensification of food production require new forms of ethical analysis (Potter, 1971). The thought that global food production should be expanded by any means necessary *can* be supported by ethical principles, but it is also countered by the need to preserve uncultivated areas for the preservation of biodiversity. In addition, principles prioritizing the rights of smallholding farmers can constrain agricultural reforms intended to intensify production systems. Moral arguments ranking the need to feed human beings above these other concerns were explicitly challenged by population ecologists such as Garrett Hardin, who argued that when the human population exceeded the carrying capacity of the global environment, the cost in lives and misery would exceed that of tempering the impulse to expand agriculture indefinitely (Hardin, 1974).

By the 1980s, the moral imperative to feed the world was being advanced to justify the industrialization of the food system, sanctioning the use of crop protection technologies known to have toxic effects on non-target species and alleged to have human health effects through contamination of groundwater and residues in food. A massive round of farm bankruptcies in the United States raised ethical questions about the social function of household farming and the structure of rural communities. Critical studies of the environmental and social impact of Green Revolution crops in less industrialized economies linked the domestic and international debates over the future of agriculture. However, Robert Zimdahl argued that the scientific staff and leadership in both public and private agricultural research exhibited an uncritical moral confidence in the justifiability of increasing yields, preventing them from even acknowledging, much less actively participating in the intricate ethical debates around these issues (Zimdahl, 2002).

Another class of problems descend from challenges to common morality itself. Peter Singer, a former student of Hare, launched a lifelong project of reexamining the ethical foundations of human relationships with animals with his book *Animal Liberation: A New Ethics for Our Treatment of Animals*. The book was originally published in 1975 and has undergone several editions, along with other updates and modifications to his philosophical position. Singer argued that just as experiences of pain or suffering should be given equal consideration without regard to race or gender, the same should hold with regard to the pain or suffering of all sentient creatures. *Animal Liberation* also included chapters documenting what Singer took to be the unethical treatment of animals in several areas, including livestock production (Singer, 1975). Soon others were arguing for the philosophically more radical idea that all vertebrate animals possess the characteristics historically thought to serve as the basis for moral rights, a view that made vegetarianism mandatory on moral grounds (Regan, 1983).

These attempts to rethink the moral standing of animals continue to provoke turmoil within the common morality, and there is little doubt that animal agriculture has felt the effects of animal liberation and animal rights. In 2004, Bernard Rollin delivered a keynote at the annual meeting of the American Society for Animal Science. Rollin described a cultural change that he characterized as 'a new social ethic' had modified public expectations for the treatment of livestock. He argued that concentrated animal feeding operations fail to pass even traditional ethical tests for ethical husbandry (Rollin, 2004). Research on livestock welfare in food production accelerated within the agricultural research establishment. As argued by David Fraser, the operationalization of this research requires scientific tools for measuring physiological indicators of welfare and estimating cognitive affect, but these measurements must be supported by ethical judgments about what matters for animal welfare (Fraser, 1999). Fraser is not talking about common morality. Animal welfare science requires disciplined philosophical reflection and exchange of views.

This brief survey of the topics calling for critically reflective agricultural ethics is far from exhaustive. If there was any topic able to shake the moral confidence of the agricultural research establishment, it was the public's reaction to the use of recombinant DNA in the development of new crop varieties. An early controversy over the animal drug recombinant bovine somatotropin sparked a congressionally mandated moratorium in the United States. Although the product was eventually approved for use in the United States, it was rejected in Europe and Canada, and by 2020, many American dairy cooperatives and milk processors had banned its use due to continuing consumer resistance. European rejection of all so-called GMOs was the most dramatic rejection of gene technologies. ('GMO' is an acronym for 'genetically modified organism' that is possibly as misleading as all agricultural plants and animals have been genetically modified.) While psychology and political activism play roles in this phenomenon, the questions such as when hesitancy is appropriate, whether labeling should be mandatory, who controls the technology and its interaction with the hunger, farmer rights and animal welfare issues discussed above all involve philosophical issues that common morality is not equipped to deal with (Thompson, 2020).

5 Methods in agricultural ethics

As the preceding sample of topics testifies, agricultural ethics is an inherently multidisciplinary form of inquiry. Virtually every social science research method can be relevant to assembling and analyzing the relevant empirical data, and many methods from the natural and computational sciences are important, as well. However, by the late 1970s, Johnson was arguing that humanistic methods were also necessary. In criticizing the agricultural sciences as overly positivistic, Johnson was calling out the agricultural disciplines' avoidance of explicitly normative methods, and he meant to include the work of agricultural economists and rural sociologists along with that of agronomists and others who applied biophysical methods (Johnson, 1976). His point can be appreciated as a feature of the root questions researchers in these fields typically ask. As a scholarly discipline, ethics is an organized inquiry in pursuit of questions like 'How should

I (or we) act?' or 'What is the right thing to do, given these circumstances?' The theoretical dimension in ethics hones in the meaning or criteria for *should* or *right*, while in other scientific disciplines, the assumed criteria for *should* and *right* are exempted from further questioning so that more effective means can be identified. The notion that this is how agricultural scientists *should* proceed is what Johnson meant by positivism.

The list of humanistic methods for this 'honing in' is too long to summarize in a book chapter, but two very general themes can be identified. First, much of the expertise in philosophical ethics (and also law) comes from mastering terminology that functions to classify and categorize normative concepts and patterns of reasoning. For example, philosophical training in ethics allows the practitioner to partition most normative evaluations into one of three categories: deontology, consequentialist (or utilitarian) and aretaic (or virtue-oriented). Deontological ethics treats rules and duties as the stopping point for an ethical inquiry and often takes some notion of rights (or justified claims on the activity of others) as their fundamental subject matter. Consequentialist theories assign value to the outcome precipitated by an action or policy and derive normativity by applying a decision rule to a comparative evaluation of likely outcomes from each of the options available to a decision maker. Utilitarianism is the most common form of consequentialist theory, urging decisions that require considering the change in welfare for all affected parties and selecting the option that maximizes welfare satisfaction. Aretaic ethics (or virtue theory) takes character or excellence to be the heart of ethical inquiry and understands normative ideas in terms of relationships that promote social virtues such as trust, confidence, citizenship and mutual affection.

This brief characterization of the three forms does not, of course, equip the reader with the ability to apply them in an ethical analysis. Expertise in ethics arises from extensive familiarization with cases, variations and nuances within each major type, much as training in law requires both facility with legal terminology and mastery of the way that judges have applied these concepts in specific cases. As discussed briefly above, some philosophers take the view that one of these major types of ethical thinking is the correct one. They may then argue that apparent applications of the other two can be reduced to some interpretation of their favored type. Others see each type as having different strengths and weaknesses and view selecting one approach, rather than another, or developing some combination of types to be a process of judgment applied in specific cases. All philosophers are sensitive to the way that positions derived from one type can create stereotypical patterns of contradiction with other types. The tension between utilitarian arguments for increasing agricultural yields at all costs and the rights of smallholding farmers is an example of such a pattern (see Kenmore et al., 2004).

The second theme concerns the method for moving from the mastery of concepts and interpretive schemes to a judgment of what is ethically right in a particular case. At the most rudimentary level, ethicists and jurists both apply analogies to compare cases on the presumption that like cases should be treated similarly. Methods for bringing the judgment of multiple minds to bear on this process include structured and semi-structured debate, where antagonists challenge one another's analysis. In law, this is done before a judge or jury, while in philosophy it is the public (including posterity) that functions as the jury. It is a lengthy, uncertain and fallible process that can take decades or more to play out, but the fact that human slavery has no defenders in the present age demonstrates that it can be decisive. More rapid, if also more tentative, results are achieved by treating the scholarly community as the jury or by combining ethical debate with participatory methods that utilize panels recruited for the express purpose of rendering a judgment (see Thompson, 2014).

Thomas Beauchamp and James Chidress developed a bioethics guidance framework for ethical reflection called *principlism*. A principlist review might convene physicians, technical specialists and representatives from patient groups with bioethicists to conduct a structured conversation on a particular treatment or technical procedure. The group works systematically through four principles, discussing how the treatment or procedure could raise ethical issues with respect to each. The four principles are as follows:

1. *Benevolence*: The benefits or goods that affected parties will enjoy as a result of the technology's implementation or use.
2. *Non-maleficence*: A principle derived from the medical principle 'do no harm'. Adverse impacts, compromises to welfare and harm experienced by affected parties as a result of the technology's implementation or use.
3. *Autonomy/dignity*: The ways in which an affected party's freedom or fate are limited or compromised by the use (or even existence) of the technology. The ways in which the technology might result in a failure to respect these beings or persons for who and what they are.
4. *Justice*: The ways in which the use of the technology or its aftermath might be considered unfair. Ethically significant discrepancies in the distribution of benefits and burdens (Beauchamp and Childress, 2019).

Professor Ben Mepham, an agricultural scientist at the University of Nottingham, modified the principlist framework so that it would be more relevant to agricultural ethics. Mepham treated the four principles listed above as column headers and then added a list of rows headed participants, third parties, non-humans and the environment, creating a four-by-four matrix. Participants in an ethical review work through all 16 cells discussing how the four principles of benevolence, non-maleficence, autonomy and justice apply to each type of affected party (Mepham, 2008). Mepham's ethical matrix is intended to function as a heuristic or rubric for collaborative reflection and discussion, as reflected in Beauchamp and Childress's use of principlism in medical rounds or IRB. The rubric is non-directive with respect to the outcome of decision-making; it merely guides collective review by highlighting areas that might be overlooked in a conventional marketing or cost–benefit analysis or omitted in a biological or engineering risk assessment. The ethical matrix has been used widely in Europe to evaluate both large-scale and small-scale innovation projects, many with applications in agriculture (see Mepham, 2000; Kaiser et al., 2007; Cotton, 2009; Webster et al., 2010; Jensen et al., 2017).

6 Agricultural ethics: why it matters

There are two general reasons why agricultural scientists and the educational and research institutions in which they work should engage in the critical, philosophically reflective activities of agricultural ethics. The first is that the outcome of debates affects the fate of their work, as well as the practice of agriculture itself. Scientists who expect their findings and innovations to be utilized can find themselves disappointed when other parties create ethically based barriers to the uptake of their research. The second is agricultural scientists should strive to do what is ethical. In cases where common morality does not provide guidance or where key tenets of common morality are being challenged, critically reflective agricultural ethics is agricultural scientist's best hope for doing the right thing. This section discusses each reason why agricultural ethics matters.

The lack of widespread public support for gene technology demonstrates what happens when scientists and research organizations fail to be full participants in the process of ethical inquiry. The controversy over genetically engineered crops and livestock species cannot be reduced to a single cause, but ethical objections to the use of rDNA gene transfer on food plants and animals were in print well before the earliest widespread plantings of herbicide-tolerant and *Bacillus thuringiensis* crops appeared in the late 1990s. A more robust ethics research capacity within agricultural universities and experiment stations would have at least prepared the science community to anticipate controversy (Thompson, 2020). More potentially contentious technologies are poised for application during the second quarter of the twenty-first century. A more thorough and better-institutionalized capacity for ethics can help leaders anticipate social challenges and potential pitfalls that these technologies might face as artificial intelligence and robotic technologies are introduced into agricultural production systems (Sparrow and Howard, 2021).

Published by Burleigh Dodds Science Publishing Limited, 2023.

The first set of reasons appeals to scientists' hopes for getting uptake from their work. In that respect, they function as a specific type of self-interest. To the extent that agricultural scientists genuinely believe that their work will serve the interests of humanity, these tactical or instrumental reasons for ethical inquiry become part of the effort to persuade others. Here, the work of ethics conjoins with the expertise of communication specialists. However, the development of a communication strategy bleeds seamlessly into the second set of reasons why ethics matters. One must ask questions like, is a factually false or insincere communications strategy offset by the benefits of a technology's eventual success? For example, Jeffery Burkhardt has argued that agricultural scientists often overstate the evidence for future benefits in order to gain public support for a particular line of research. A thoroughgoing utilitarian might conclude that stretching the truth is justified by the benefits of convincing others to support the position that leads to the best outcome. However, as Burkhardt notes, neither philosophers nor the public has understood the duty to speak truthfully in such terms (Burkhardt, 2001). One can argue that science has such a strong alliance with norms of truth-telling that deliberate perpetuation of falsehoods or oversimplifications is forbidden by the deontological (i.e. rule-based) norms that govern the very process of science. With this thought, the explanation of why ethics matters shifts from the role of persuasion into the second category of reasons: supporting scientists' responsibility to act ethically *tout court*.

In considering what it means for scientists to act ethically, it is helpful to recall Hare's distinction between common morality and the deliberative thinking associated with philosophical ethics. Common morality applies to everyone. It dictates norms of honesty and non-maleficence that fall on scientists just as they fall on everyone. It provides the rationale for condemnation of egregious instances of scientific fraud (Kennedy, 2006). However, there are aspects of the scientific enterprise that test the applicability and justifiability of these norms, leading to the need for more sophisticated tools of philosophical analysis. The most dramatic examples have been outside the agricultural sciences. Nazi doctors conducted painful experiments resulting in death, disfigurement and mental torment of prisoners in concentration camps. In the United States, a study designed to track the long-term development of syphilis in a group of black men was allowed to continue even after penicillin was known to cure the disease. In both cases, scientists believed that the long-term collective benefit to humankind outweighed the costs to individuals who were victimized by these studies (Berger, 1990; Brandt, 1978).

The Nazi experiments and the Tuskegee study are now regarded as paradigm examples of unethical research. There is thus a sense in which the norms requiring informed consent from research subjects have been absorbed into common morality. This was not clear at the time the offending science was done, not so long ago. The claim that benefits would outweigh harm was plausible and supported by a particular understanding of utilitarian ethics. The utilitarian argument was eventually overturned by the deontological principle requiring informed consent. These cases from the history of biomedical research exemplify the sense in which unreflective adherence to a single ethical principle can lead scientists into practices that, over time, are considered to be deeply unethical. Beauchamp and Childress's priciplism was intended as a rubric that would help scientists avoid such errors, though, as discussed earlier, the creation of IRB oversight may have undercut the reflective dimension of ethical review, converting the process into a bureaucratic exercise in checking tick boxes (Rollin, 2006). The point to stress is that reflective and open-minded review (e.g. the process of doing ethics) can and should be instrumental in making the process of scientific research ethical.

In fact, though, there is more. Scientists stand in a fiduciary relationship with members of the public that extends their responsibilities beyond that of common morality. This means that scientists have a responsibility to use their expertise in a way that is consistent with the interests of their clients, the public at large. In their study on the social responsibilities of gene drive research, Pamela Sankar and Margaret Cho review successive editions of the US National Research Council publication *On Being a Scientist*, as well as documents on scientific responsibilities from the US National Institutes of Health and various Presidential bioethics commissions. Although these sources state a responsibility to serve the public, Sankar and Cho find their statements on social responsibility to be vague. In some cases, they suggest that scientists have fulfilled their responsibilities simply by policing fraud and observing IRB requirements, then doing 'good science' – a notion defined in terms of rigor, rather than moral responsibility. However, they note that gene drive researchers have gone further, foreswearing activities that, though profitable for researchers and their funders, could involve applications that are not consistent with the public interest (Sankar and Cho, 2015). This is the essence of fiduciary responsibility, a term used in the financial industry to distinguish financial advisers that serve client interests from those who adopt a 'buyer beware' attitude toward investment products on offer.

In other contexts, I have questioned whether scientists *do* adhere to the ethics of fiduciary responsibility. It is not clear, for example, that scientists working in the private sector put the public interest ahead of opportunities for financially lucrative research (Thompson, 2018). Although there may be circumstances in which research tools are legitimately deployed in pursuit of self-regarding ends, on my reading, *On Being a Scientist* describes the role of the scientist qua scientist as implying an overriding concern for the public welfare that would not be associated with a typical for-profit venture. The third edition includes sections on how scientists should approach the potential for conflicts of commitment, as well as more conventional conflicts of interest, and encouraging scientists to subject their ideas to procedures of collective assessment (NRC, 2009). The fiduciary responsibility in agricultural science is arguably stronger still, as practitioners in universities often cite 'the land grant mission' of conducting research that is in the public interest. Attention to ethics in the sense of a deliberate and critically aware process of reflective evaluation is central to the land grant mission (Zimdahl, 2003).

7 Can agricultural ethics be taught?

Whether or not agricultural ethics is teachable depends on what one means by agricultural ethics and what one means by teaching. In this section teaching will be understood as a formalized activity conducted in venues such as classes, workshops and short courses. It is distinguished from mentoring, advising and the overly broad 'teaching by example'. I take it as obvious that ethics of all kinds are indeed taught by example, but since one seldom knows when one is serving as an example, this form of schooling is not amenable to discrete, planned activities designed specifically to bring about a well-defined end. We commonly speak of our entire life's experience as contributing to our education, and in this sense, it is obvious that ethics is something that we learn. This does not exclude the possibility that some people learn it better than others. However, to the extent that ethics becomes a learning objective to be addressed by definite pedagogical techniques, it is less clear that agricultural ethics is teachable.

In these closing remarks, I rely on some 40 years of experience in classroom instruction, as well as individual instruction, to offer some generalizations. The background to this is, again, Hare's distinction between common morality and philosophically reflective ethics, which includes agricultural ethics. Although common morality is definitely learned, it is not really taught. Common morality is an accumulation of habit and socialization into categories for mutual engagement with other people and the world at large. Any attempt to shift from these largely unconscious processes of acculturation toward a didactic presentation of content is either propaganda or already a step in the philosophical direction. The primary result of classes intended to teach principles of common morality is to reinforce the notion that somebody in a position of authority over the curriculum thinks it is important. This is not insignificant, but it is nonetheless unlikely to result in an edificatory intellectual process.

Agricultural ethics, on the other hand, can be taught, but only if ethics is understood as an activity, as something that people do, rather than as the codified result of such a process, like a set of rules or an algorithmic decision procedure. Robert Zimdahl's career-crowning book, *Agricultural Ethics – An Invitation* covers a much more exhaustive set of topics than have been reviewed in this chapter. It was intended as a book that might be used in an advanced undergraduate or even graduate ethics class for the agricultural sciences (Zimdahl, 2020). Yet even if Zimdahl's book is teachable, I cannot imagine teaching from it in the manner that the bulk of the content for agricultural science education is taught. In my experience, the best-structured learning occurs when students are presented with contrasting views on a topic and encouraged to analyze each view, laying out the strong and weak points in each position. On the one hand, a summary of competing arguments is itself an analysis of a controversy and does not substitute for a student's reading and original analysis of primary materials. On the other hand, such summaries can be enormously helpful to people who are trying to make sense of a disagreement. However, such people are not really students; they are already participants in an ongoing ethical debate.

This observation points toward an underappreciated aspect of ethical pedagogy. The *timing* of ethics education is crucial. The educational psychologist Lawrence Kohlberg (1927–1987) argued that children apply different approaches to moral thinking at different stages in their intellectual development. They proceed from simply avoiding punishment through forms of conventionalism before reaching the ability to weigh the costs and benefits in their late teens and early 20s. Kohlberg thought that most people would not reach the most advanced stage, a form of self-regulating deontology in which people understand the right action in terms of universalizable duties until approximately their mid-20s (1981). The details of Kohlberg's theory are contested, especially in light of empirical research showing that women appear to take a rather different path, but my point is simply to note that even if he is only vaguely correct, people are not in a position to undertake sophisticated moral reasoning during the years that they are generally undergoing classroom education. Agricultural ethics is more likely to have an impact when it is conducted as an integral component of the research process, as opposed to something relegated to the curriculum.

The theme that runs throughout this chapter is for readers to be mindful of the difference between the agricultural ethics that is something we can expect everyone to have and something that functions as an activity, as something that people - especially agricultural scientists - do. I have characterized the former as a common morality, a set of norms and standards for practice that people (should) internalize through social interactions with others. There are times when this process goes astray, and there are sociopaths for whom it fails altogether. It is reasonable to think that a program or center on agricultural ethics would take up some of the social psychology and communications research needed to address such problems, but that is not how agricultural ethics has been defined for the purposes of this chapter. Agricultural ethics is a specific form of reflective, deliberative multidisciplinary inquiry that combines elements from each discipline in the agricultural sciences (including agricultural economics and rural sociology) with humanistic methods from history, law, literature and, of course, philosophy. It will be most effective when focused on specific, controversial issues. It can help researchers and policymakers avoid gross pitfalls and misunderstandings in the pursuit of their activities, but more importantly, it can help them get things right at the start.

8 Conclusion

When one hears complaints about the lack of ethics among farmers, food industry firms or other actors in the food system, it is usually an individual or incorporated group's failure to abide by the terms of common morality. Sometimes these are moral failures on the part of the actors in question. Someone or some institutionally organized person has failed to abide by commonly accepted norms. They have allowed laxity or some personal interest to override commonly accepted norms of honesty and non-maleficence. I have argued that while philosophers can help us recognize the *nature* of these failures (that is what I have been doing in this chapter), calling in a philosopher is probably not the most effective way to correct such problems. However, there are also failures that arise because individuals face conflicting social expectations, and where changes in infrastructure, technology or policy make it difficult to fulfill expectations without making unreasonable personal sacrifices. There are also cases where expectations themselves are changing, and that change can also affect what we take a failure to be. It is in these more complex examples of moral failure that philosophical ethics can help.

The potential for confusion over this difference between failures in common morality and challenges that call for a structured, critical reflection creates a barrier to the institutionalization of ethics within agriculture. People who are focused on common morality are right to be skeptical about the benefits of research or teaching in agricultural ethics. This barrier is further reinforced by disciplinary norms within the agricultural sciences, including economics, sociology and other social sciences. Of course, there are also actors in the food system who have disreputable reasons to avoid scrutiny. These factors discourage the type of inquiry that Robert Zimdahl has urged upon his colleagues in the biologically oriented agricultural sciences.

Yet, agriculture has seen two centuries of dramatic transformation in its technical base. While contemporary production systems produce high yields, they are manifestly plagued by environmental problems, problems that will only be amplified in the era of rapid climate change. What is more, longstanding injustices in agricultural labor have not been resolved, while the rate of diet-based disease climbs. At the dawn of the twenty-first century, unprecedented amounts of venture capital are being invested in the agricultural sector. Innovators seek to marry machine learning, sensing technology and big data to the chemical, genetic and mechanical technologies that dominated the nineteenth and twentieth centuries. Posterity deserves a reflective questioning that evaluates the potential - positive *and* negative - implicit within these unprecedented changes. That is what agricultural ethics is, and why it matters.

9 Where to look for further information

Several of the sources cited in the reference section provide more detailed or alternative approaches to the definition and application of agriculture. Readers should consult Mepham, 2008 and Beauchamp and Childress, 2019 as general introductions to general methods in bioethics. Johnson, 1976 and Rollin, 2006 discuss challenges for conducting ethical inquiries in the context of scientific research. Robert Zimdahl's two books (2002 and 2020) provide a more extended introduction with discussion of specific examples and case studies of relevance to agricultural research and the practice of farmers and ranchers. There are dozens of introductory overviews covering basic concepts in philosophical ethics. Many scientists have found *The Elements of Moral Philosophy* by an American philosopher, James Rachels (1941-2003) to be especially useful. The book continues to be released in new editions with contributions from Rachels' son Stuart.

Peer-reviewed journals such as *The Journal of Agricultural and Environmental Ethics* and *Agriculture and Human Values* are the primary home for topical studies in agricultural ethics. Both are indexed in AGRICOLA, the U.S. National Agricultural Library's electronic search tool. In general, work on specific topics in agriculture can be carried out with an electronic search of scholarly journals by conjoining the word *ethics* to the topic of interest (e.g. search on 'agricultural biotechnology' and 'ethics'). However, some science-oriented indexes, such as PubMed, do not index peer-reviewed journals in the humanities and some social science fields, and many important studies in agricultural ethics continue to be published in monographs. A less accurate but more inclusive tool such as Scholar Google may produce more satisfactory results. In addition, Springer publishing's *Encyclopedia of Food and Agricultural Ethics* is a continuously updated electronic resource.

The primary professional organization for agricultural ethics is the European Society for Agriculture and Food Ethics (https://www.eursafe.org/), along with the subsidiary Asia-Pacific Society for Agriculture and Food Ethics (https://www.apsafe.online/). The North American scholarly community is more diffuse, but many agricultural ethicists participate in the Agriculture, Food and Human Values Society (https://afhvs.wildapricot.org/).

10 References

Beauchamp, T. L. and Childress, J. F. (2019). *The Principles of Biomedical Ethics* (8th edn.). New York: Oxford University Press.

Berger, R. L. (1990). Nazi science–The Dachau hypothermia experiments. *New England Journal of Medicine* 322(20): 1435–1440.

Brandt, A. M. (1978). Racism and research: The case of the Tuskegee Syphilis study. *Hastings Center Report* 8(6): 21–29.

Burkhardt, J. (2001). Agricultural biotechnology and the future benefits argument. *Journal of Agricultural and Environmental Ethics* 14(2): 135–145.

Cotton, M. (2009). Evaluating the ethical matrix as a radioactive waste management deliberative decision-support tool. *Environmental Values* 18(2): 153–176.

Davies, C. (2020). Tesco blackmail trial: Farmer convicted of spiking baby food. *The Guardian*. Accessed August 20. Accessed October 31, 2021. Available at: https://www.theguardian.com/business/2020/aug/20/tesco-blackmail-trial-farmer-nigel-wright-convicted-spiking-baby-food-blackmail-plot.

Fraser, D. (1999). Animal ethics and animal welfare science: Bridging the two cultures. *Applied Animal Behaviour Science* 65(3): 171–189.

Hardin, G. (1974). Lifeboat ethics: The case against helping the poor. *Psychology Today Magazine* 8: 38–43, 123–126.

Hare, R. M. (1981). *Moral Thinking: Its Levels, Method, and Point*. New York: Oxford University Press.

Jensen, K. K., Michalopoulos, T., Meijboom, F. L. B. and Gjerris, M. (2017). Perceptions of ethical challenges within the low input breeds project. *Food Ethics* 1(2): 109-125.

Johnson, G. L. (1976). Philosophic foundations: Problems, knowledge, solutions. *European Review of Agricultural Economics* 3(2–3): 207-234.

Kaiser, M., Millar, K., Thorstensen, E. and Tomkins, S. (2007). Developing the ethical matrix as a decision support framework: GM fish as a case study. *Journal of Agricultural and Environmental Ethics* 20(1): 65–80.

Kenmore, P. E., Stannard, C. and Thompson, P. B. (2004). *The Ethics of Sustainable Agricultural Intensification*. Rome: Food and Agricultural Organization.

Kennedy, D. (2006). Responding to fraud. *Science* 314(5804): 1353.

Kohlberg, L. (1981). *The Philosophy of Moral Development: Moral Stages and the Idea of Justice*. New York: Harper & Row.

Marosi, R. (2014). Hardship on Mexico's farms, a bounty for U.S. tables. *The Los Angeles Times*. Accessed January 1, 2022. Available at: https://graphics.latimes.com/product-of-mexico-camps/.

Mepham, B. (2000). A framework for the ethical analysis of novel foods: the ethical matrix. *Journal of Agricultural and Environmental Ethics* 12(2): 165–176.

Mepham, B. (2008). *Bioethics: an Introduction for the Biosciences* (2nd edn.). New York: Oxford University Press.

NRC (National Research Council). (2009). *On Being a Scientist: A Guide to Responsible Conduct in Research*. Washington, DC: National Academies Press.

Potter, V. R. (1971). *Bioethics: Bridge to the Future*. Englewood Cliffs, NJ: Prentice-Hall.

Regan, T. (1983). *The Case for Animal Rights*. Berkeley, CA: University of California Press.

Rollin, B. E. (2004). Annual meeting keynote address: Animal agriculture and emerging social ethics for animals. *Journal of Animal Science* 82(3): 955-964.

Rollin, B. E. (2006). *Science and Ethics*. New York: Cambridge University Press.

Sankar, P. L. and Cho, M. K. (2015). Engineering values into genetic engineering: a proposed analytic framework for scientific social responsibility. *The American Journal of Bioethics* 15(12): 18–24.

Sharma, K. and Paradkar, M. (2010). The melamine adulteration scandal. *Food Security* 2(1): 97–107.

Singer, P. (1975). Animal liberation: A ethic for our treatment of animals. In: *New York Review*. New York: Random House.

Sparrow, R. and Howard, M. (2021). Robots in agriculture: Prospects, impacts, ethics and policy. *Precision Agriculture* 22(3): 818–833.

ten Have, H. A. (2012). Potter's notion of bioethics. *Kennedy Institute of Ethics Journal* 22(1): 59–82.

Thompson, P. B. (2004). The legacy of positivism and the role of ethics in the agricultural sciences. In: Scanes, C. G. and Miranowski, J. A. (Eds.), *Perspectives in World Food and Agriculture 2004*. Ames, IA: Iowa State University Press, pp. 335-351.

Thompson, P. B. (2014). Agricultural ethics. In: Thompson, P. B. and Kaplan, D. M. (Eds.), *Encyclopedia of Food and Agricultural Ethics*. New York: Springer Reference, pp. 54–61.

Thompson, P. B. (2018). The role of ethics in gene drive research and governance. *Journal of Responsible Innovation* 5(Suppl 1): S159–S179.

Thompson, P. B. (2020). *Food and Agricultural Biotechnology in Ethical Perspective* (3rd edn.). New York: Springer.

Webster, J., Bollen, P., Grimm, H., Jennings, M. and Steering Group of the RETHINK Project. (2010). Ethical implications of using the minipig in regulatory toxicology studies. *Journal of Pharmacological and Toxicological Methods* 62(3): 160–166.

Zimdahl, R. L. (2002). Moral confidence in agriculture. *American Journal of Alternative Agriculture* 17: 44–53.

Zimdahl, R. L. (2003). The mission of land grant colleges of agriculture. *American Journal of Alternative Agriculture* 18(2): 103–115.

Zimdahl, R. L. (2020). *Agricultural Ethics: An Invitation*. New York: Springer.

Chapter 2

Approaches to ethics

Katie McShane, Colorado State University, USA

1 Introduction

Ethics is the study of moral rightness, wrongness, goodness, and badness. It concerns questions of what to do, what to care about, and how to live well in the world together with others. Ethics is a subfield of philosophy and is sometimes referred to as 'moral philosophy'. (The terms 'ethical' and 'moral' are typically used interchangeably in philosophy.) Ethics uses the methodology of philosophy, which differs in significant ways from the methodology of the sciences, whether natural or social. The methodology of philosophy centers on the presentation and evaluation of arguments; such arguments may contain conceptual claims (about the relations among concepts), normative claims (prescriptive or evaluative claims), and empirical claims (about evidence gleaned from observation and experimentation).

There are three main branches of ethics, and each branch asks different questions about moral rightness, wrongness, goodness, and badness. The first branch is metaethics, which asks about the nature of moral judgments in general. Central to metaethics are questions about what one is doing when one makes a moral judgment. When I say, 'Lying is wrong', am I stating a fact, expressing a feeling, prescribing a course of action, or doing something else? Metaethics also asks questions about the metaphysical assumptions of ethics. Is value something external – something 'out there' in the world that we perceive – or is it an aspect of our own subjective states?

http://dx.doi.org/10.19103/AS.2023.0125.02

The second branch of ethics is normative ethics. Normative ethics is sometimes simply referred to simply as 'ethics'; it is typically the subject of Introduction to Ethics courses at universities. Normative ethics asks which things in the world are right, wrong, good, and bad and what makes them that way. For example, many people think that it is wrong to cheat or to cause others to suffer unnecessarily. Normative ethics offers theories about why these things are wrong - about which feature(s) of these behaviors makes them wrong.

The third branch of ethics is applied ethics, sometimes referred to as practical ethics. The traditional approach to applied ethics is to take theories from normative ethics and apply them to practical problems in the world in order to determine the ethical solution to those problems. For example, if it is wrong to cause unnecessary suffering to others, then ethics might forbid certain experiments that are extremely painful for research subjects. However, sometimes the standard theories in normative ethics are not so easily applied to practical problems, particularly those involving new technologies or unprecedented situations. For example, standard ethical theories do not have much to say about whether it would be wrong to clone human beings. For this reason, many applied ethicists reject the top-down approach traditionally taken by applied ethics and instead advocate for a more reciprocal critical relationship between normative and applied ethics: using normative ethical theories to help solve practical problems, but also using the complications posed by practical problems as a reason for revising and improving theories in normative ethics. Applied ethics encompasses a broad range of topics: bioethics, legal ethics, military ethics, business ethics, political ethics, computing and artificial intelligence ethics, animal ethics, environmental ethics, and agricultural ethics, among others.

This chapter focuses on contemporary normative ethical theories that come out of Western philosophical traditions and on secular rather than religious ethical theories since these approaches predominate in Western countries and international institutions. In explicating these theories, this chapter will pay particular attention to the way these theories have been applied to - and have been shaped by - work in animal, environmental, and agricultural ethics. (See Chapters 1, 3, 4, and 5 in this book for a more detailed explication of agricultural ethics.)

In Western philosophical traditions, there are three main schools of thought within normative ethics: utilitarianism, deontology, and virtue ethics. Each school of thought takes a different aspect of ethics as its starting point. The result is three theories that tell different stories about what it is to be ethical, though in practice these theories often recommend and forbid the same actions.

One distinction cross-cuts these theories and is of particular importance to animal, environmental, and agricultural ethics. This is the distinction between anthropocentric and nonanthropocentric ethical theories. Anthropocentrism means 'human-centeredness'. Anthropocentric ethical theories consider only humans and human welfare to be directly important to ethics; any things that are not human (animals, plants, species, ecosystems) matter only for their effect on humans and their interests; they do not matter in their own right. Nonanthropocentrism rejects this limitation of the scope of ethics to human beings. Depending on the theory in question, anthropocentrism can take different forms. All three types of theory - utilitarian, deontological, and virtue ethical - have both anthropocentric and nonanthropocentric versions. In considering the three schools of thought below, the differences between anthropocentric and nonanthropocentric versions of the theories will be noted.

2 Utilitarianism

The first school of thought in normative ethics is utilitarianism. Utilitarianism is a consequentialist theory, which means that it evaluates the rightness or wrongness of an action by looking at the goodness or badness of the action's consequences. The right course of action to take, according to a consequentialist, is the course of action that produces the best consequences. As consequentialists, utilitarians believe that the right course of action is the one that produces the best consequences; they claim further that the best consequences are those that involve the highest level of overall well-being. The utilitarian slogan, to paraphrase Jeremy Bentham, is 'the greatest good for the greatest number' (Bentham, 1776/1969: 45).

Utilitarians require some way of measuring overall well-being in order to assess which course of action produces the highest level of it. In classical utilitarianism, overall well-being is simply the total of the well-being of all individuals. While this might seem obvious, it has two implications worth noting. First, the good of a society for a utilitarian is nothing more than the good of each of its constituent members added up. Utilitarians thus disagree with philosophers such as Rousseau (1762/1968), who believed that there is a 'common good' that is not reducible in this simple way to the good of individuals. Second, for a utilitarian, the well-being of each individual matters, and each individual's well-being is counted in the same way toward the sum total. Thus the well-being of a king or a chief executive officer counts for neither more nor less than the well-being of a subsistence farmer or a refugee. What matters is the level of well-being, not whose well-being it is. The utilitarian goal is to make choices that produce the highest level of total well-being in the world - that is to say, the greatest total benefit and the least total harm - regardless of whom those benefits or harms accrue to.

There is some disagreement within utilitarianism about how to understand what well-being (also referred to as welfare, utility, or interests) really is. A theory about what well-being is (in the literature, often referred to as a theory of welfare) is necessary if one is to measure the levels of well-being. The earliest utilitarians, Jeremy Bentham (1789/1948) and John Stuart Mill (1863/1979), believed that well-being was simply a matter of happiness or suffering. On this view, sometimes referred to as hedonistic utilitarianism, the best course of action is the one that produces the greatest amount of happiness in the world and the least amount of suffering.

Other utilitarians see well-being as a matter of the satisfaction of desires. According to this view, someone's level of well-being is a matter of whether their desires are satisfied, i.e. whether they get what they want. According to this type of utilitarianism (usually referred to as preference- or desire-satisfaction utilitarianism), the best course of action is the one that satisfies the most desires and frustrates the fewest desires. Desire-satisfaction utilitarianism has been subject to widespread criticism in philosophy, resulting in the development of a variant: rational-desire-satisfaction utilitarianism (for discussion, see Griffin (1986)). Critics point out that the satisfaction of one's desires is not always good for one, notably in cases where the desires have been formed on the basis of mistaken beliefs. For example, if I think that getting 200 mg of vitamin A in my diet every day will improve my health, I might desire to consume 200 mg of vitamin A each day. But clearly doing this would not increase my level of well-being, since such an amount would be toxic. Rational-desire-satisfaction utilitarians claim instead that well-being is a matter of the satisfaction of one's rational desires - desires not formed on the basis of false beliefs or other types of mistake - and thus that the best course of action is the one that satisfies the most rational desires and frustrates the fewest rational desires (see, e.g. Railton (1986)).

Finally, there are some theories that regard well-being as a matter of the satisfaction of basic needs or the development of basic capabilities (see, e.g. Nussbaum (2000); Kraut (2007); Sen (1993)). In the philosophical literature, this type of view is referred to as perfectionism, and its historical roots go back to ancient Greek philosophy. According to this view, a person's level of well-being is a matter of whether their basic human needs (food, water, shelter etc.) are met and whether their basic human capabilities (for communication and affiliation, language and rational thought, decision-making and autonomy etc.) are developed.

It is common to find utilitarian ethical theories deployed in large-scale policy assessment, particularly cost-benefit analysis. Since the ultimate utilitarian goal is to improve overall well-being as much as possible, utilitarians care about assessing the effect that different policy choices would have on the well-being of all those affected by them. Cost-benefit analysis also aims to assess the costs and benefits of different policy choices. In practice, however, cost-benefit analyses tend not to assess the effects of policies on the level of overall well-being in the world, which is what utilitarians are concerned with, but rather a more limited set of costs and benefits. Non-economic costs and benefits, and costs or benefits to future generations, are often not well represented in cost-benefit analyses, which has been a source of criticism (see, e.g. UNFCCC (2013)). It is worth noting that because the historical roots of utilitarianism are also those of economics, utilitarianism and economics deploy similar methodologies – and even use some of the same terminologies. However, one must be careful: the two fields have developed in different directions, and so superficial similarities can sometimes be misleading. The terms 'utility' and 'value', for example, have taken on different meanings in each field.

All utilitarians share the view that the right thing to do is whatever course of action leads to the highest level of overall well-being. Anthropocentric utilitarians, however, assume that 'overall well-being' refers only to the well-being of humans, whereas nonanthropocentric utilitarians believe that 'overall well-being' includes the well-being of nonhumans – such as animals – too. Historically, many utilitarians were anthropocentrists, either explicitly or implicitly; they viewed ethics as a question of how people ought to treat other people and so only considered the well-being of people in their theorizing. There are many challenges to anthropocentrism within utilitarianism, however, both historically and more robustly in contemporary theorizing (see, e.g. Singer (1990), discussing Bentham (1789/1948: p. 310, n. 1)). At least some animals can suffer; they have desires; and they have basic needs and capabilities. Thus at least some animals have a well-being of exactly the kind that utilitarianism is concerned with. Because of this fact, and because of utilitarianism's commitment to treating each individual's well-being as mattering regardless of the identity of the individual to whom the well-being belongs, contemporary utilitarians typically see no principled reason for excluding animal well-being from the sphere of ethical concern. Because of this, concerns about animal welfare (as opposed to animal rights; see below) have often been part of utilitarian thinking about the environment. However, the place of plants and nonsentient animals, as well as ecological wholes such as species and ecosystems, has been more precarious in utilitarian thinking. The scientific consensus thus far is that plants do not feel suffering, nor do simpler animals, nor ecological wholes such as species and ecosystems. These entities also do not seem to have desires at all, much less rational desires. Thus hedonist, desire-satisfaction, and rational-desire-satisfaction utilitarians do not consider them to have a well-being that is part of the overall well-being that they aim to improve. Arguably at least some of these entities do have basic needs and capabilities, however, and so there might be room for greater inclusion of such entities within a perfectionist utilitarianism. To date, however, few theories of this kind have been developed (for an exception, see Fulfer (2013)).

To understand how utilitarianism works in practice, consider a problem such as climate change. In thinking about how to respond to this problem, utilitarians focus on the effects that climate change will have on the well-being of all those who are or could be affected by it. This includes the effects not only on one's own family, or the people in one's region or industry, or even the people in one's own country: it includes all people in the world, as well as all future generations of people, and any nonhumans (current or future) that one's version of utilitarianism includes. A utilitarian will determine which responses to climate change are possible and then recommend the one that produces the greatest overall level of well-being, i.e. the greatest benefit and the least harm for all affected. This is a broader view than many decision-makers and policy advocates typically take. In the formation of climate policy, for example, many politicians focus only on the effects that climate change will have on their country, state, or province; many industry groups focus only on the effects that climate change will have on their industry. According to utilitarianism, an ethical approach to climate change must take a broader view than this: it must consider the good of all, not just the good of some.

3 Deontology

The second school of thought in normative ethics is deontology, which is an approach to ethics that focuses on duties, and in many cases on rights associated with those duties. Unlike consequentialist theories such as utilitarianism, deontology does not evaluate the ethical acceptability of an action by looking at the consequences produced by the action. Rather, it looks at the kind of action it is. A deontological ethic consists of certain rules that require or prohibit actions of certain kinds. In general, deontological theories involve articulating rules of conduct, giving a justification for those rules, and explaining what to do in cases of conflicts (or apparent conflicts) among the rules. Immanuel Kant is probably the best-known theorist of deontological ethics. Kant (1785/1996) claimed that it is wrong to lie or cheat, for doing so violates rules that one expects others to follow. A successful lie, Kant noted, requires a background of trust. If everyone lied, nobody would believe what anyone says, and so one's attempt to deceive someone with a lie wouldn't work. Likewise, cheating requires the existence of practices or institutions in which most other people follow the rules. If all students cheated on their exams in school, for example, teachers would stop using exams to assess students' learning. In order to cheat successfully, then, there must be enough people who follow the rules that these institutions and practices remain intact. Kant also claimed that one has a duty not to manipulate, exploit, or coerce people; rather, one should treat all rational agents (i.e. beings capable of acting for reasons), including oneself, with respect. Treating rational agents with respect means not interfering with their autonomy – their ability to guide their own lives by the choices they make according to reasons they endorse. Interferences with autonomy include deception, manipulation, exploitation, and coercion. These are all ways of interfering with another person's reasoning process in order to get it to produce the result you want or preventing their decisions from effectively guiding their lives.

While Kant thought that all of these requirements were ultimately contained in one all-encompassing moral principle, which he called the Categorical Imperative, W.D. Ross (1930/2002), a later deontologist, argued instead that we have many different duties, and in any given case, they might come into conflict with one another. In virtue of this difference, Kant is considered a monistic deontologist (claiming a single, over-arching duty), while Ross is considered a pluralistic deontologist (claiming many duties, not reducible to one another). Ross argued that one has a duty to keep one's promises, to right one's wrongs, to exhibit gratitude and try to return favors, to promote the good of others, to avoid harming others, to try to improve oneself, and to treat others fairly.

Theories of rights are also a part of many deontological theories. Rights are simply entitlements to be treated or not treated in a certain way. Ethical theories discuss moral rights – moral entitlements to treatment – as opposed to legal rights, which are entitlements that one has only because one lives under some legal system or another. Sometimes moral rights are cited to explain moral duties, such as the ones described above. For example, a deontologist might claim that we have a duty not to lie to others because doing so would violate their right to autonomy. This focus on rights has made deontological theories particularly useful as a foundation for political theories, which tend to centrally involve claims about rights. Deontological ethical theories are prominent in areas of social life where rules, entitlements, and prohibitions are centrally important: in medical ethics, military ethics, professional codes of conduct, and international human rights, to name a few. In these areas, deontologists sometimes argue that their theories offer a better guide to action than utilitarian theories: they tell people simply not to do certain things (e.g. kill a healthy person and distribute their organs to people awaiting transplants; torture prisoners of war; falsify data) rather than telling them to decide whether doing those things would have better consequences than not doing them.

Deontological theories have historically been anthropocentric, but in the early 1980s nonanthropocentric versions were developed, most notably within animal ethics (see, e.g. Regan (1983)). Animal rights theories, as opposed to animal welfare theories (which have roots in utilitarianism), focus on animals' rights to be treated or not treated in certain ways. Just as Kant argued that it is wrong to treat other humans as mere resources for the achievement of one's own satisfactions, for example by manipulating or exploiting people, some animal rights theorists argue that it is wrong for humans to treat animals as mere resources: creatures that exist merely to feed us, do work for us, or entertain us. Animal rights theorists have offered different views about what rights animals have: a right to life and a right to self-defense are common claims; other theorists have argued for extending property rights or political rights to animals as well (see, e.g. Donaldson and Kymlicka (2011) and Hadley (2005)). While a few theories have tried to extend rights beyond humans and animals (e.g. Taylor (1981)), such theories are still relatively rare in ethics.

In assessing a problem such as climate change, deontologists focus on the duties that different parties have to one another and the rights that - according to some theories - ground these duties. For example, the duty to right one's wrongs might suggest that those who have been most responsible for causing climate change should accept the greatest responsibility for alleviating it (Gardiner, 2006). The duty to avoid harming others might suggest that there is a moral duty to reduce one's greenhouse gas emissions. The duty to keep one's promises might suggest that it is not only politically unwise but also ethically wrong for countries to fail to abide by international agreements requiring mitigation measures.

In considering climate risks and harms, a deontologist's focus on rights might lead to a somewhat narrower scope of analysis than the utilitarian's. Rather than ask which policies would lead to the greatest level of overall well-being now and in the future, a rights-oriented deontologist will focus more closely on aspects of climate change that would involve rights violations (Caney, 2010). While both the loss of life and reduced corporate profits are costs produced by climate change, for example, a deontologist might regard the loss of life as a violation of a basic human right (the right to life), but not the reduction of corporate profits. In considering how to assess and manage risk due to climate change, then, the deontologist would prioritize risks to life over risks to corporate profits.

4 Virtue ethics

The third school of thought in normative ethics is virtue ethics. Virtue ethics has a long history in Western philosophical traditions, going back at least as far as the ancient Greeks. Virtue ethical theories focus on what it is to be a virtuous - that is to say, morally good - person (see Chapter 5). They assess behaviors in terms of the character traits that those behaviors express: courage, honesty, cruelty, etc. Virtue ethical theories typically contain a list of virtues and vices, an explanation of what makes a trait a virtue or a vice, a description of each virtue and vice, and an explanation of the extent to which the virtues and vices vary according to context. In Western philosophy, the views of Aristotle (350 BCE/1984) have probably had the greatest influence on contemporary virtue ethics. Aristotle described many virtues, among them courage, temperance, generosity, truthfulness, patience, and wit. He believed that for most of the virtues, the virtuous state is the mean between two extreme states and that each of the extreme states is a vice. For example, courage is the mean between the extreme of rashness on one side and cowardice on the other. When it comes to showing confidence in the face of fear, rashness is the vice of showing too much confidence (e.g. walking across a busy street with complete confidence that you won't get hit by a car). Cowardice is the vice of showing too little confidence (e.g. never leaving the house because you might catch a cold). Courage is showing just the right amount of confidence: neither too little nor too much. This view is often referred to as the Doctrine of the Mean or the Golden Mean.

Aristotle's virtue ethics assesses not only people's actions and choices but also their feelings. He famously described virtue as a matter of having the right feelings 'at the right times, with reference to the right objects, towards the right people, with the right aim, and in the right way' (Aristotle, 350 BCE/1984: p. 1747, 1106b21-22). Virtuous people do not have to force themselves to do the right thing; they want to do the right thing; likewise, they do not have to stop themselves from doing the wrong thing; they don't want to do it in the first place. Aristotle also thought that virtue ethics requires organizing societies in ways that make the virtues easier to develop and express.

Virtue ethical theories have historically been anthropocentric; accounts of virtues and vices have mainly been described in terms of the ways that humans interact with one another. However, many virtues and vices apply just as well to the ways that humans behave toward nonhuman parts of the world. One can be generous, compassionate, or cruel to nonhuman animals; one can express respect or arrogance in one's treatment of other species or even ecosystems. The concepts of virtue and vice have long been applied nonanthropocentrically in 'nature writing' (see, e.g. Leopold (1970), Carson (1962), and Bilbro (2019)). In more recent years, the development of 'environmental virtue ethics' has provided a philosophical framework for nonanthropocentric ascriptions of virtue and vice (see., e.g. Hill (1983), Sandler (2007), and Treanor (2014)).

In contemporary ethics, virtue ethical theories have been influential in the philosophy of education, and in certain areas of professional ethics: social work, nursing, and teaching most prominently. Virtue ethical theories have also been used to critique moral norms on a larger scale. For example, needlessly high rates of consumption in Western countries have been tied to social norms that celebrate the vice of greed and downplay the virtues of temperance and modesty (see, e.g. Sagoff (2001)).

When assessing the problem of climate change, virtue ethicists ask which virtues or vices are expressed by different behaviors. For example, some argue that both climate change itself and attempts to fix it technologically (space mirrors, sulfate injection, and the like) are hubristic - as expressing a kind of human arrogance about humans' place in the natural world (see, e.g. Preston (2013: p. 26)). Others assess the world's failure to prevent the problem as a matter of selfishness, callousness, or willful ignorance (see., e.g. Williston (2015)).

5 Competing theories or an ethical toolkit?

Utilitarianism, deontology, and virtue ethics clearly focus on different aspects of ethical life, but are these competing theories about how to be ethical, or are they all different parts of the same story? Most philosophers regard them as competing accounts since they offer different theories about what is fundamental to ethics, and they provide different – and largely incompatible – justifications for their theories. On the other hand, many people who want to use these theories to help them make ethically good decisions would prefer to treat the three theories as part of a single overall account of ethics. Which view should one take? On the one hand, philosophers are not wrong that there is significant disagreement among the theories. One cannot be a Benthamite, a Kantian, and an Aristotelian in any even minimally consistent way. For example, both Aristotle and Kant argue against thinking of a good human life only in terms of the amount of pleasure it contains, a central assumption of Bentham's ethics (Aristotle, 350 BCE/1984: pp. 1852–1864, 1172a19–1179a32; Kant, 1785/1996: pp. 49–50, Ak. 4:393-394). Aristotle and Bentham argue against the kind of absolute rules that are the foundation of Kant's ethics (Aristotle, 350 BCE/1984: pp. 1747–1797, 1106a14–1138b17; Bentham, 1789/1948: pp. 1–23). While contemporary versions of utilitarianism, deontology, and virtue ethics each try to capture the central insights of the other approaches, there is still a considerable amount of disagreement. Within philosophy, attempts to combine theories in some principled way typically end up treating one theory as foundational. For example, indirect utilitarianism might direct us to follow deontological rules because doing so produces the highest level of overall well-being (for discussion, see Gray (1984); Gray (1996)).

Someone who prefers to view the three approaches as an 'ethical toolkit', however, might argue that ethical life has different aspects, and each school of thought in normative ethics does particularly good job of theorizing one of them. Wanting to be ethical involves wanting to make the world a better place, wanting to do the right thing, and wanting to be a good person. Utilitarianism has a particularly well-developed theory of what it would be for the world to be a better place: it is for the creatures in the world to fare as well as possible. Deontology has a particularly well-developed theory of right and wrong action: it is wrong to lie, murder, exploit, manipulate, etc.; it is right to help others, to keep one's promises, to return favors, etc. Virtue ethics has a particularly well-developed theory of what it is to be a good or bad person: to be a good person is to be honest, courageous, generous, etc. and to be a bad person is to be rash, petty, deceitful, cruel, etc. Perhaps for this reason, many ethical guidelines combine the values, rules, and ideals from each of the three schools of thought described above into their articulations of what it is to be ethical. The AMA's code of ethics (American Medical Association, 2022), for example, requires that physicians work toward the betterment of the community (utilitarianism), respect the rights of patients (deontology), and exhibit compassion (virtue ethics). The NIH's Principles of Ethical Conduct (National Institutes of Health, 2022) includes rules prohibiting certain actions, such as accepting gifts from entities seeking action from the NIH (deontology), and it requires that employees act in ways that express particular virtues, such as loyalty, honesty, and impartiality (virtue ethics). The NSF's Ethical Code (National Science Foundation, 2022) discusses contributions to welfare, such as maintaining a healthy workplace and protecting the environment (utilitarianism); duties and rights, such as reporting workplace discrimination and respecting human rights (deontology); and virtues, such as respectfulness and professionalism (virtue ethics). These ways of bringing together the insights about ethical behavior from different theoretical sources needn't be problematic, so long as ethical codes or boards have a coherent view about how they fit together – including how to prioritize them and what to do in cases of conflict. In practice, the 'toolkit' approach is made easier by the fact that utilitarianism, deontology, and virtue ethics often agree about what one should or shouldn't do in particular cases, even though they may each offer different reasons to explain why. For example, all three theories will tell researchers not to falsify data. Utilitarianism will justify this by pointing to the bad consequences produced by falsifying data; deontology will justify it by pointing to the principle that one ought not to lie; and virtue ethics will justify it by noting that it expresses the vice of dishonesty.

6 Conclusion

Utilitarianism, deontology, and virtue ethics are three alternative accounts of what makes actions right, wrong, good, or bad. While each offers a different theoretical framework and focuses on different aspects of moral behavior, each account provides a helpful set of guidelines for ethically good behavior.

7 Where to look for further information

Good general introductions to ethics:

- Driver, J. (2007). *Ethics: The Fundamentals*. Malden, MA: Blackwell.
- Kagan, S. (1998). *Normative Ethics*. Boulder, CO: Westview Press.
- Shafer-Landau, R. (2020). *The Fundamentals of Ethics,* 5th ed. Oxford: Oxford University Press.

Key journals in ethics:

- *Ethical Theory and Moral Practice*.
- *Ethics*.
- *Philosophy and Public Affairs*.
- *Utilitas*.

8 References

American Medical Association. (2022). AMA code of medical ethics. Available at: https://www.ama-assn.org/sites/ama-assn.org/files/corp/media-browser/principles-of-medical-ethics.pdf (accessed September 2, 2022).

Aristotle. (1984). Nicomachean ethics. In: Barnes, J. (Ed.), *The Complete Works of Aristotle*, vol. 2. Princeton, NJ: Princeton University Press, pp. 1729-1867.

Bentham, J. (1776/1969). A fragment on government. In: Mack, M. P. (Ed.), *A Bentham Reader*. New York: Pegasus, pp. 45-72.

Bentham, J. (1789/1948). *The Principles of Morals and Legislation*. New York: Hafner Press.

Bilbro, J. (2019). *Virtues of Renewal: Wendell Barry's Sustainable Forms*. Lexington: University of Kentucky Press.

Caney, S. (2010). Climate change, human rights, and moral thresholds. In: Gardiner, S. M., Caney, S., Jamieson, D., et al. (Eds.), *Climate Ethics: Essential Readings*. Oxford: Oxford University Press, pp. 163–177.

Carson, R. (1962). *Silent Spring*. New York: Houghton Mifflin.

Donaldson, S. and Kymlicka, W. (2011). *Zoopolis: A Political Theory of Animal Rights*. Oxford: Oxford University Press.

Fulfer, K. (2013). The capabilities approach to justice and the flourishing of nonsentient life. *Ethics and the Environment* 18(1): 19–42.

Gardiner, S. M. (2006). A perfect moral storm: Climate change, intergenerational ethics and the problem of moral corruption. *Environmental Values* 15(3): 397–413.

Gray, J. (1984). Indirect utility and fundamental rights. *Social Philosophy and Policy* 1(2): 73–91.

Gray, J. (1996). *Mill on Liberty: A Defence*. London: Routledge.

Griffin, J. (1986). *Well-Being: Its Meaning, Measurement, and Moral Importance*. Oxford: Clarendon Press.

Hadley, J. (2005). Nonhuman animal property: Reconciling environmentalism and animal rights. *Journal of Social Philosophy* 36(3): 305–315.

Hill, T. E. (1983). Ideals of human excellences and preserving natural environments. *Environmental Ethics* 5(3): 211–224.

Kant, I. (1785/1996). Groundwork of the metaphysics of morals. In: Gregor, M. J. (Ed.), *Practical Philosophy*. Cambridge: Cambridge University Press, pp. 39–108.

Kraut, R. (2007). *What is Good and Why: The Ethics of Well-Being*. Cambridge, MA: Harvard University Press.

Leopold, A. (1970). *A Sand County Almanac With Essays on Conservation From Round River*. New York: Ballantine Books.

Mill, J. S. (1863/1979). *Utilitarianism*. Indianapolis: Hackett.

National Institutes of Health (2022). Principles of ethical conduct for government officers and employees. Available at: https://ethics.od.nih.gov/principles-ethical-conduct-government-officers-andemployees (accessed September 2, 2022).

National Science Foundation. (2022). Living our values: NSF code of ethics. Available at: https://nsfinternational.widen.net/s/fwvwkvwcgj/nsf_code-of-ethics_a4_en_lco272 (accessed September 2, 2022).

Nussbaum, M. C. (2000). *Women and Human Development: The Capabilities Approach*. Cambridge: Cambridge University Press.

Preston, C. J. (2013). Ethics and geoengineering: Reviewing the moral issues raised by solar radiation management and carbon dioxide removal. *WIREs Climate Change* 4(1): 23–37.

Railton, P. (1986). Moral realism. *Philosophical Review* 95(2): 163–207.

Regan, T. (1983). *The Case for Animal Rights*. Berkeley, CA: University of California Press.

Ross, W. D. (1930/2002). *The Right and the Good*. Oxford: Oxford University Press.

Rousseau, J.-J. (1762/1968). *On the Social Contract*. New York: Penguin Books.

Sagoff, M. (2001). Consumption. In: Jamieson, D. (Ed.), *A Companion to Environmental Philosophy*. Oxford: Blackwell Publishing, pp. 473–485.

Sandler, R. L. (2007). *Character and Environment*. New York: Columbia University Press.

Sen, A. (1993). Capability and well-being. In: Nussbaum, M. C. and Sen, A. (Eds.), *The Quality of Life*. Oxford: Oxford University Press, pp. 30–53.

Singer, P. (1990). *Animal Liberation*. New York: Avon Books.

Taylor, P. W. (1981). The ethics of respect for nature. *Environmental Ethics* 3(3): 197–218.

Treanor, B. (2014). *Emplotting Virtue: A Narrative Approach to Environmental Virtue Ethics*. Albany, NY: SUNY Press.

UNFCCC. (2013). *Non-Economic Losses in the Context of the Work Programme on Loss and Damage: Technical Paper*.

Williston, B. (2015). *The Anthropocene Project: Virtue in the Age of Climate Change*. Oxford: Oxford University Press.

Chapter 3

Institutionalizing agricultural ethics in US land-grant universities

Robert L. Zimdahl, Colorado State University, USA

We can, of course, be deceived in many ways. We can be deceived by believing what is not true; but we certainly are also deceived by not believing what is true.

S. Kierkegaard (1995) – Works of Love.

1 Introduction

When something is institutionalized it is established as a convention or norm of an organization or culture. Most professional disciplines have published and institutionalized professional ethical standards. Universities routinely include study of ethics in the curriculum for medicine, law, business, engineering, and the environment. The lack of thought about and consideration of agriculture's ethical standards is of concern because agriculture, the essential industry, lacks a defined moral foundation. Study and consideration of the standards governing the conduct of members of the agricultural profession that determine the principles of right and wrong have not been institutionalized in US Land Grant University Colleges of Agriculture. Nor have they been promulgated by agriculture's professional organizations, government agencies, or in the many parts of the agribusiness industry. The objective of this chapter is to offer reasons why institutionalization has not occurred and suggest consequences and solutions.

http://dx.doi.org/10.19103/AS.2023.0125.03

2 Food production and 'productivism'

The ethical position that characterizes agriculture is 'productivism'. It is the central, indeed often the only norm. The moral imperative is to produce food and fiber to benefit all humanity. Production is what must be sustained and expanded. Those involved in agriculture whether they are producers, suppliers, or researchers and regardless of their employer should ask and debate if maintaining and increasing food production is a sufficient criterion for judging the consequences of all agricultural activities. Does increasing production justify everything agriculture does? Does it achieve sustainable production practices, contribute to social justice, and reduce agriculture's effects on climate change? Does the quest to increase production solve or even address agriculture's moral dilemmas?

Agricultural scientists have assumed that as long as their research and the resultant technology increased food production and availability, they and the end users were somehow exempt from negotiating the moral bargain that is the foundation of the modern democratic state (Thompson, 1989). It is unquestionably a moral good to feed people. Therefore, it is assumed, anyone who questions agriculture's morality and the uses and results of its technology simply doesn't understand the importance of what is done and why it is done. It is assumed that agricultural practitioners are technically capable and the good results of their technology make them morally correct. We are obliged to question that assumption as Wendell Berry (1977) has.

> We have lived by the assumption that what was good for us would be good for the world. We have been wrong. For I do not doubt that it is only on the condition of humility and reverence before the world that our species will be able to remain in it.

When those involved in agriculture claim credit for improving production and keeping food cost low, they must also accept society's right to hold them responsible for problems they have regarded as externalities[1]. They need to ask and be prepared to respond to what has not been asked often enough - what could go wrong? What has gone wrong?

3 Ethical standards

There are explicit ethical standards among people in agriculture. All have a sense of what is right and wrong, which is often unexamined and not supported by careful reasoning. A guide toward helping decide what one ought to do is

[1] An externality is a cost that is not reflected in price, or more technically, a cost or benefit for which no market mechanism exists. It is a loss or gain in the welfare of one party resulting from an activity of another party, without there being any compensation for the loss. From a self-interested view, an externality is a secondary cost or benefit that does not affect the decision-maker. For example, dealing with pollution from nitrogen fertilizer is not the responsibility of the user or the manufacturer of the fertilizer. It is a public/governmental responsibility.

found in our societies' principles. They are guides to help one decide what is good or bad, moral or immoral, and right or wrong. We are subject to social ethical expectations about torture, pornography, civil rights, women's rights, and treatment of children. Scientists have professional ethics: don't fabricate data, give proper credit, be honest, include opposing results, and disclose conflicts of interest.

I suggest the truest test of the mission of a scientific discipline, indeed of one's life, is to examine its ethical base and the reasons for it. It is 'an exercise in reason to find where the weight of reason rests' (Rachels and Rachels, 2007). Many assume agriculture has had and now has an adequate ethical foundation. This assumption has not been questioned. Moral questions have been explored in natural resources and environmental studies because public pressure demanded it. Moral evolution occurs because ethical principles, once articulated, examined, and found lacking, are abandoned in favor of other more carefully constructed thinking. Our respective cultures provide several examples of change: child labor, smoking in restaurants, women's rights, environmental rights, and animal rights. But there has been no comparable change and little critical thinking about agriculture's moral foundation.

4 Scientific challenges in agriculture

Agriculture has scientific challenges such as maintaining production, minimizing its effects on climate change, dealing with pesticide resistance, loss of biodiversity, invasive species, biotech/GMOs, and concern about and achieving sustainability. Many involved in agriculture believe development and use of more and more energy-dependent technology has been good, and is required to assure food production, and will solve the problems caused by the unintended consequences of current and future technologies.

I do not mean to imply that we should abandon science and technology. We humans, the Earth's dominant species, are not just figures in the landscape - we are shapers of the landscape (Bronowski, 1973, p. 19). Having achieved this power we should think carefully about whether our shaping of the landscape is desirable and sustainable. Most of those involved in agriculture have a clear idea of what they intend to do. They should think about and consider what their actions may be undoing. They must develop in themselves, their colleagues, and their students' critical thinking about the moral dimension of what they do and undo.

Agriculture is our most widespread interaction with the environment. We live in a post-industrial, information-age society. No one will ever live in a post-agricultural society. Continuing to justify all agricultural activities and technology by the necessity of achieving the moral obligation and production challenge of feeding a growing world population has not been and will not be a sufficient defense for agriculture's negative environmental and human effects. Humans are no longer just a part of nature; we are a force of nature that is disturbing and changing the climate and our planet's ecosystems at a pace and scope never seen in human history (Friedman, 2016): 'We cannot rebuild the Greenland ice sheet, the Amazon rain forest, or the Great Barrier reef and Koalas. When the macaws, the rhinos, and the orangutans are gone, no 3D printer will bring them back to life.' The social, environmental, and economic costs of a developed country's capital, energy, and chemically dependent agricultural system, and the challenge of sustaining the environment and other species are recognized by some in the agricultural community as important, but increasing production dominates the agricultural realm.

5 Problems facing agriculture

What problems have resulted from agriculture's ecosystem effects and the emphasis on production? The world now produces enough food to feed everyone an adequate daily diet. Feeding the 11 billion people expected to be on the planet at the end of this century is undeniably a good thing. Is it a production problem? Of course it is. But more than enough food is produced now to feed the global population. Nevertheless, 811+ million people still go hungry every day[2]. After steadily declining for a decade, world hunger is on the rise, affecting 1 in 9 of the world's people. From 2019 to 2020, the number of undernourished people grew by as many as 161 million, a crisis driven largely by international and local conflicts, climate change, and the COVID-19 pandemic. In spite of the abundance of food, people are hungry because of poor food distribution by government agencies, inadequate infrastructure for food distribution, food storage loss, waste by consumers[3], governmental policies, and poverty. More production alone will not solve the hunger problem. Sen (1999, p. 162) confirmed that undernourishment, starvation, and famine are influenced by the working of the entire economy and society – not just by food production and agricultural activities. Poverty, hunger, and famine are rarely caused or solved by more food production.

6 Agriculture's ethical dilemmas

It is obvious to anyone who listens to, reads, or watches the news that citizens of democratic societies are becoming increasingly reluctant to entrust their water, diets, and natural resources blindly into the hands of farmers, agribusiness firms, and agricultural scientists. Ethicists and agricultural practitioners must

[2] Sustainable Development Goal2, zero hunger:https://sustainabledevelopment.un.org/sdg2. Accessed June 2020.
[3] In the US more than one-third of available food is never eaten. https://www.usda.gov/foodlossandwaste. Accessed August 2022.

participate in a dialog that leads to social consensus about the positive and negative effects of agriculture's technology, its risks, and reasonable solutions. In the past most risks were borne by users of technology. Now there is widespread concern that the risks and short- and long-term consequences of agricultural technology are borne by others. There is important, rational concern about pesticides in our food and the environment, the role and future effects of genetic modification, deforestation of the Amazon to grow soybeans and pasture cattle, cruelty to animals, harm to migrant labor, and agriculture's contributions to climate change. Agriculturalists must begin to contribute the time and resources needed to listen and understand their positions and those of their fellow citizens. All involved in agriculture must recognize they are responsible for how we ought to live. For most non-agricultural segments of society, this is not a new demand. For agriculture it is.

Many people throughout the world, in both developed and developing countries, have concerns about the ethical dimensions of agriculture and our food system that go beyond the central need to feed humanity. Each of agriculture's multiple responsibilities includes an ethical dimension:

Achieving sustainability - What must be sustained? Who decides?

Surface and groundwater - Irrigation is essential to successful agriculture. Who will decide how the world's diminishing supply of this finite resource will be allocated?

Biotechnology/GMOs - Genetic modification is viewed by many agricultural scientists as a major contributor to increasing food production. However, there is great concern and fear among the public supported by vague theories, hopes, and some public figures about genetic modification.

Climate change - How the agricultural realm deals with climate change is and will remain a major responsibility.

Exploitation and inhumane treatment of farm labor - Migrant labor is essential to the growth of many important crops. The laborers are not treated well and often work and live under inhumane conditions.

Harm to other species - Throughout the world species are going extinct. There is abundant information on the negative effects of agriculture on a variety of habitats

Cruelty to animals - Animal rights have become a major issue in the United States and other countries. It is clear animals can suffer and should not.

Loss of crop genetic diversity - Narrowing of the genetic diversity of crops is a continuing concern.

Loss of small farms and rural communities - Small farms (200 acres +/−) are disappearing across the United States. A consequence is small villages and rural culture disappear.

Pollution of water, soil, and humans from pesticides, fertilizers, and animal waste.

The nutritional value of foods provided to consumers by the food system. Is a rising concern.

The power of corporate farming and its lack of transparency These and surely other issues require all engaged in agriculture to discuss and work together toward mutually agreeable solutions.

7 An ethical perspective on agriculture

These are not just scientific problems. We should not expect scientists to solve them although scientific research and new technology will be required. It is reasonable to hope some segments of the agricultural enterprise will work together with others to mitigate/reduce agriculture's negative effects and its ethical dilemmas. All segments of the agricultural enterprise must cooperate to achieve worthy, morally good goals. Agriculture will gain little if it wins the production battle and loses the moral battle

The way agriculture is practiced, research projects are chosen and conducted, and the teaching done involves scientific and ethical values. From an ethical perspective, feeding the growing world population is clearly a morally good thing, but it does not absolve the agricultural community from critical, ethical examination of the totality of agriculture's effects.

Within universities, colleges of agriculture have emphasized what to do and think rather than how to think. The faculty and administrators of colleges of agriculture have not pulled away from addressing ethical issues - they have regarded them as technical, scientific issues or ignored them. Agriculture's ethical issues have not been ignored by others (see Blatz, 1991; Caparròs, 2015; Jackson, 1980; LeMay, 1998; Thompson et al., 1994; Thompson, 1995, 1998; Wojick, 1989). *The Journal of Environmental Ethics* began in 1978, and the *Journal of Agricultural and Environmental Ethics* began in 1988. Both journals have discussed agriculture, but few authors have been from colleges of agriculture. Universities have traditionally been places where different opinions were welcomed and encouraged. The present trend in public education toward specifying what controversial topics may or may not be taught is disturbing. Teaching about controversial topics is now forbidden or avoided because teachers fear being accused of offending someone's beliefs. This stands in sharp contrast to the role of teaching - to lead out - to educate. Encouraging students and the general public to be aware of, and discuss difficult, controversial issues is an important role of education and those who teach. Discussion of such issues is never easy.

What can our universities do to institutionalize agricultural ethics, albeit a controversial topic? A place to begin is the classroom. The agricultural curriculum lacks courses in agricultural ethics that focus on general ethical principles and their application to agricultural issues. In 2012 such courses were available at only nine US Land Grant universities with agricultural colleges (Zimdahl and Holtzer, 2016), having declined from 15 in 1999 (Zimdahl, 2000). A 2022 survey of all 50 US land-grant universities with colleges of agriculture showed:

- Forty-six offer a class on environmental ethics in the Department of Philosophy. Many included agriculture's effects.
- Seven offered environmental ethics in the department of philosophy and in another department.
- Two have an environmental ethics class in another department but not in philosophy.
- Two did not offer a class on environmental ethics.

The 2022 survey included only the 50 original land-grant universities. I hope someone will expand the study to other institutions. Two examples:

- Environmental ethics including agriculture is taught in the Department of Philosophy at Middle Tennessee State University in Murfreesboro, TN.
- Agricultural and Natural Resource Ethics is taught in the Department of Food and Resource Economics at the University of Florida.

A more inclusive survey might reveal related content in several other institutions

- In 2022, agricultural ethics is offered by only five land-grant universities (University of Connecticut, University of Georgia, Iowa State University, University of Pennsylvania, and University of Vermont). I suggest the paucity of university courses on agricultural ethics at most US land-grant universities exists because
- The faculty who plan the curriculum and advise undergraduate and graduate students do not and have not regarded studying the ethical values of agriculture as important preparation for agricultural professionals.
- As a student, I was not and my professors and their mentors were not advised to enroll in any classes in philosophy during their undergraduate and graduate education.
- Agricultural faculty are not interested in cooperating with a colleague in the department of philosophy to create a class on agricultural ethics and encouraging students to enroll. When I created and taught the agricultural ethics class at Colorado State University most of the students were from the Department of Philosophy or a department in the College of Liberal Arts while only a few were from the college of agriculture. Because budgets of land-grant universities have tightened over the last several decades,

reducing or denying support for an agricultural ethics class is an easy option in an already crowded curriculum.
- Few agriculture students are advised to enroll in any classes in the department of philosophy. They might, but in most universities they cannot enroll in an agricultural ethics class that does not exist.

Creating and encouraging students to enroll in such classes will not alone quickly increase the emphasis on agricultural ethics (Zimdahl and Holtzer, 2018). It will be a recognition of the need to acknowledge and discuss agriculture's ethical dimensions. Agriculture has economic problems, which have focused attention on production and profit while education and practice have ignored agriculture's human and ethical dilemmas (Damasio, 1999). I suggest professors, department heads, and deans of colleges of agriculture who have chosen not to address agriculture's ethical dilemmas are contributing to present and future problems. There is a clash between the environmental and human harm of modern agricultural production, the values held by the general society, and those who practice agriculture in different ways. The risk of ignoring the value conflicts and societal concerns will lead to a further loss of public support and confidence in agriculture.

The academic data and recommendations above and conversations with colleagues and friends lead me to believe agriculture's technology may outweigh the character of those involved in our food system. We have not risen to the level of our aspirations, we hold at the level of our training - our education. We risk approaching what Niebhur (1932) meant when he accused his colleagues of being moral people in an immoral profession. J. S. Mill (1859, Chap. 2) warned us - 'He who knows only his side of the case knows little of that.' Mill urged us to seek out conflicting views of our side of the case. Are parts of agriculture immoral? Discussions and reliable answers - a cure for bias - will be found in interaction with people who don't share our beliefs and who confront us with counter-evidence and counterarguments (Haidt, 2022). Acknowledging agriculture's moral dilemmas and institutionalizing their study may transform us and the practice of agriculture.

We, and others, are a mass audience consuming the same content while looking in a mirror that reflects the view we have (Haidt, 2022). We are constrained by a 'collective egoism compounded by the egoistic impulses of individuals which achieve a more vivid expression and have a more enduring effect when they are united rather than expressed separately and discreetly' (Niebhur, 1932, p. 25, introduction 1960 Ed.). The university serves a unique role in our collective conscience. It can be a place that brings us together to discuss difficult topics in a respectful, edifying way. Rather than being an ideological echo chamber - as social media so often become - it should be a place where varying views of the world (agricultural and other) are discussed rigorously, kindly, toward the goal of making the world a more habitable place for all things. The role of university is critical to addressing agriculture's ethical dilemmas.

Science cannot answer the inevitable questions when two or more opposing moral questions arise. When the morally good goal of feeding a growing world population bumps up against the morally good goal of protecting the environment one is confronted with value questions that science is not designed to and cannot answer. When the environment's natural objects are valued only in terms of their worth to humans they can be and are legally destroyed or modified. Christopher Stone (1972, p. 10) was among the first to ask if natural objects have rights - a radical concept. Kolbert (2022) returned to Stone's question and notes 'for most of history people saw themselves as dependent on the environment.' Kolbert correctly notes that it now seems normal to send in the bulldozers, chainsaws, and backhoes to cut down the trees, fill the wetlands, and 'develop' the environment.

Our agricultural practices have contributed to changing the climate, acidification of the oceans, and pollution. Little, if any, attention is paid to the inevitable environmental consequences: climate change, ocean hypoxic areas, soil erosion, melting ice, species extinction, and invasive species. Our predatory self-interest dominates our environmental concern. Until the environment and its many components receives a right we often see them only as things for our use. The objection that streams and forests cannot speak was, in Stone's (p. 3) view, easily addressed. Neither corporations, American states, estates, infants, incompetents, municipalities nor universities can speak. Yet, as Stone reminds us these entities are amply represented in the courts. We make decisions on behalf of and in the purported interest of others every day. The other creatures whose wants are far less verifiable may be more important. They are more metaphysical in conception than the wants of rivers, trees, and land and the human obligation to them (Kolbert).

Niebhur (1932, p. 3) asked 'if it's possible for human intelligence to increase the range of benevolent impulses and encourage us to consider the needs and rights of others and other things (Nash, 1977) in addition to the things to which we are bound by organic and physical relationships.' Can we transcend our own interest and grant rights to the interests of our fellow humans and the many creatures in the environment? If agriculture's practitioners continue to ignore agriculture's moral dilemmas because we must produce, they may lose the ability and right to determine agriculture's future and jeopardize our collective chances of surviving on this planet (Berry, 1977). If we fail to institutionalize the study of the ethics of agriculture we will not learn how to ask and discuss moral questions. We should not continue to defend only the interests of agriculture when there are obviously unjust effects on the interests of the social community.

Human ingenuity has increased the treasures nature provides for the satisfaction of human needs, it will never be sufficient to satisfy all human wants.

If all elements of the agricultural enterprise, including farmers/ranchers/producers, agribusiness firms, and food processors and sellers, do not begin to recognize and address agriculture's ethical dilemmas there will be several enduring effects. Three bad outcomes seem most likely.

First. Agriculture practitioners may find their arguments and justifications for their technology and production practices ignored by the society and decision-makers or their relevance and legitimacy may simply be denied.

Second. Public unease, dissatisfaction, and fear of agricultural technology, products (e.g. pesticides, cruelty to animals, food quality, climate change), and production technology will result in increasing societal unrest and pressure for political action. Decisions on how agriculture is to be practiced and how land is to be treated will be made by society, government agencies, and legislators. Such decisions will not be aided by thoughtful discussion - the hallmark of university education.

Third. All involved in agriculture will miss opportunities to improve agriculture's ability to address its multiple responsibilities and its negative effects will worsen. The increasing concentration of food production in the hands of agribusiness companies will continue. Small farms, farmers, and rural communities will continue to gradually disappear.

Action by the agricultural community to address its ethical dilemmas and impose ethical standards will reduce and avoid negative effects. It will enable the agricultural community to establish its own ethical standards rather than waiting to see if appropriate levels of sustainability and resilience can be achieved by the present capital, chemical, and energy-intensive agricultural system. If the agricultural community fails to act, ethical standards will likely be imposed by others or not established at all.

Because agriculture is a diverse widespread enterprise, reaching an agreement will be difficult, but not impossible. Recognizing possible, undesirable outcomes and choosing to act wisely will assure the continued success of the essential industry.

8 Acknowledgements

I thank Dr. James Boyd, Professor Emeritus, Department of Philosophy, Colorado State University, Fort Collins, CO; Dr. Wade Casey, Fort Collins. CO; Ms. Jennifer Galt, Washington, DC; Dr. Thomas Holzer, Professor Emeritus,

Department of Agricultural Biology, Colorado State University, Fort Collins, CO; Dr. Clarence Swanton, University Professor Emeritus, Department of Plant Agriculture, University of Guelph, Ontario, Canada, and Dr. Donald Viney, Professor of Philosophy, Pittsburg State University, Pittsburgh, KS for reviewing and improving the manuscript.

9 References

Berry, W. 1977. *The Unsettling of America: Culture and Agriculture*. Avon Books, New York, p. 228.

Blatz, C. V. 1991. Ethics and Agriculture – An anthology on current issues in world context. University of Idaho Press, Moscow, ID, p. 64.

Bronowski, J. 1973. *The Ascent of Man*. Little, Brown and Company, Boston, MA, p. 448.

Caparròs, M. 2015. *Hunger the Oldest Problem*. Melville House, Brooklyn, NY, p. 525.

Damasio, A. 1999. *The Feeling of What Happens: Body and Emotion in Making of Consciousness: A Harvest Book*. Harcourt Publishers Inc., New York, p. 285.

Friedman, T. L. 2016. *Thank You for Being Late - An Optimist's Guide to Thriving in the Age of Accelerations*. Farrar, Straus and Giroux, New York, p. 486.

Haidt, J. 2022. Why the past 10 years of American life have been uniquely stupid – It's not just a phase. *The Atlantic Monthly*, May, pp. 54–66.

Jackson, W. 1980. *New Roots for Agriculture*. University of Nebraska Press, Lincoln, NE, p. 150.

Kierkegaard, S. 1995. *Loves Hidden Life and Its Recognizability by Its Fruits. P-5 in 1847 Works of Love.* Original Publication. Princeton University Press, Princeton, NJ, p. 525.

Kolbert, E. 2022. Testing the waters - Should the natural world have rights? *The New Yorker*, April 18, pp. 16–19.

LeMay, B. W. J. 1998. Science, ethics and food. *Papers and Proceedings of a Colloquium Organized by the Smithsonian Institution*. Smithsonian Institution Press, Washington, DC, p. 143.

Mill, J. S. 1859. *On Liberty* (Vol. 1865). Longman's, Green, and Co., London.

Nash, R. 1977. *Do Rocks Have Rights? The Center Magazine* (Vol. 7). Center for the Study of Democratic Institutions, Santa Barbara, CA, pp. 2-12.

Niebhur, R. 1932. *Moral Man in an Immoral Society: A Study in Ethics and Politics*. Westminster John Knox Press, Louisville, KY, p. 284.

Rachels, J. and Rachels, S. 2007. *The Elements of Moral Philosophy* (5th edn.). McGraw Hill, New York, p. 209.

Sen, A. 1999. *Development as Freedom*. A. A. Knopf, New York, p. 366.

Stone, C. 1972. *Should Trees Have Standing? Toward Legal Rights for Natural Objects*. William Kaufman, Inc., Los Altos, CA, p. 101.

Thompson, P. B. 1989. Values and food production. *Journal of Agricultural Ethics* 2(3): 209-223.

Thompson, P. B. 1995. *The Spirit of the Soil - Agriculture and Environmental Ethics*. Routledge, New York, p. 196.

Thompson, P. B. 1998. *Agricultural Ethics - Research, Teaching and Public Policy*. Iowa State University Press, Ames, IA, p. 196.

Thompson, P. B., Matthew, R. J. and van Ravenswaay, E. O. 1994. *Ethics, Public Policy and Agriculture*. McMillan Published Company, New York, p. 268.

Wojick, J. 1989. *The Arguments of Agriculture - A Casebook of Contemporary Agricultural Controversy*. Purdue University Press, West Lafayette, IN, p. 123.

Zimdahl, R. L. 2000. Teaching agricultural ethics. *Journal of Agricultural and Environmental Ethics* 13: 229–247.

Zimdahl, R. L. and Holtzer, T. O. 2016. The ethical values in the U.S. agriculture and food system. *Journal of Agricultural and Environmental Ethics* 31(6): 751–753. https://doi.org/10.1007/s10806-018-9753-4.

Chapter 4

Gender dimensions of agricultural ethics

Samantha Noll, Washington State University, USA

1 Introduction

Agriculture is one of the oldest and most transformative of human activities, with wide-ranging social effects. Innovation in food production has been continuous throughout history, as humans learned to cultivate crops, domesticate animals, and build complicated systems of irrigation (Thompson and Noll, 2015). Today, we typically associate agriculture with the production of food crops, such as grains, fruits, and vegetables, but the term also captures a diversity of other activities, including fiber production, animal husbandry, dairy production, and aquaculture. Additional activities found under the umbrella of agriculture include the cultivation, harvesting, processing, storage, and distribution of commodities produced. When taken together, these activities constitute an important economic sector domestically and globally. For example, according to the United States Department of Agriculture (USDA), the agricultural sector contributes approximately 1.055 trillion to the U.S. gross domestic project (or GDP) in 2020 (USDA, 2017). In Canada, the contribution to GDP is around CAD134.9 billion (Government of Canada, 2022). Thus, agriculture can be understood as a bundle of activities necessary for human and economic flourishing, irrespective of context.

When viewed from a macro level, it is easy to separate agriculture from the people who perform the tasks discussed above. Agriculture, writ large, is understood as food reserves, commodity bundles, export profits, gross

http://dx.doi.org/10.19103/AS.2023.0125.04

domestic product, etc. But at the community level, agricultural production helps to ensure that families have access to culturally appropriate foodstuffs and individuals receive the calories they need to thrive. Additionally, these activities are an integral part of rural communities, as farming is a common profession globally in rural contexts (World Bank, 2014). At the level of the individual, agriculture can be understood as a way of life, where an individual works to cultivate, harvest, process, store, and/or distribute specific products. These labors often occur in a specific locality that might have deep cultural and personal value, such as on a multi-generational family farm (Whatmore, 1991; Ramey, 2014). As agriculture is made up of activities performed by people in social contexts, cultural and social structures greatly influence the lived experiences of producers.

Feminist scholars argue that social frameworks often include a gendered component that deeply influences personal identity, individual behaviors, and daily life (McAfee, 2018). Research on gender and agriculture is part of a larger body of literature that begins with attention to gender and how social roles are influenced by social and cultural roles associated with a specific sexual identity. As people who identify as women have been historically marginalized in unique ways, feminist scholarship often begins from the position of the oppressed. McAfee (2018) argues that gender-focused 'philosophical scholarship begins with attention to women, to their roles and locations. What are women doing? What social/political locations are they part of or excluded from? How do their activities compare to those of men? How have their roles been valued or devalued?' (NP). Gender-sensitive work in agriculture is no different. It also begins by asking these types of questions within the specific context of agricultural praxis. This chapter provides a detailed overview of the insights this work gleaned since its initial development in the 1970s and 1980s. While women contribute substantially to agricultural production, there are unique challenges and barriers that effect women in the agricultural sector.

Specifically, this chapter provides an overview of research on gender in agriculture, with an eye toward agricultural ethics. First, it will briefly outline the history of agricultural ethics as well as current themes in the discipline. The potential range of topics in this area is extraordinary. For example, a common theme found in popular work involves critiques of pesticides on human and wildlife populations, as inputs may have harmful effects when improperly stored, handled, or applied. Other topics include social effects to family farms and rural communities, as technological developments may lead to the consolidation of properties into larger operations (Sassenrath et al., 2008). It will then provide a detailed analysis of the 'gender turn' in agricultural ethics, beginning in the 1980s with the work of Carolyn E. Sachs (1983), who produced one of the first feminist studies of agriculture. This literature highlights unique challenges faced by female producers around the globe. For instance, in Southern Asia, sub-Saharan Africa, and Oceania, agriculture employs approximately 60% of women, though this percentage could increase to 80%-90%, depending on the country and region (Nyasimi and Huyer, 2017). Women in these rural areas are often responsible for producing crops needed to ensure local food security. Yet, in a breach of distributive justice, women are also less likely to have access to resources needed to increase productivity and are 10% more likely to be food insecure (United Nations, 2019). The chapter will end by discussing emerging themes in the literature. However, before providing an overview of research on gender and agricultural ethics, it is important to place this discussion in context.

2 Current agricultural demographics

According to the Food and Agriculture Organization (FAO) of the United Nations, nearly half of the global population (or approximately 3.1 billion people) live in rural areas and over half are women (Glazebrook et al., 2020). Of the 3.1 billion rural population, nearly 2.5 billion people list agriculture as their livelihood or as a means of subsistence (FAO, 2022, 2012). These numbers are approximations. However, the FAO concludes that 1.4 billion women rely on agriculture in some form to meet their daily needs, as well as their household caloric requirements. For example, in the global South as a whole, women make up 43% of the workforce and constitute nearly 66% of livestock keepers (FAO, 2022). These numbers highlight how women are active in agriculture globally. However, 500 million female farmers do not own the land they cultivate. Women, as a whole, typically receive substantially less agricultural resources than their male counterparts (FAO, 2012). According to Glazebrook et al. (2020), 'women contribute substantially to national food-baskets... and their agricultural productivity is crucial for meeting food security. The productivity of at least 500 million women is, however, hampered by limited access to land and resources' (pg.1). Thus, women not only play an integral part in agricultural projection globally but also face unique challenges.

The numbers above suggest that demographics in agriculture may differ substantially in the global North and South. Indeed, fewer women are engaged in agriculture in the global North. However, this is changing. According to recent census data, 'developed' countries are experiencing a resurgence of female farmers. For example, Europe has experienced a sharp rise in the number of women farmers over the last 30 years (European Commission, 2019). According to Glazebrook et al. (2020), 'by 2016, approximately 30% of farms were managed by women across the European Union (EU), though this statistic hides significant variations between countries in that Eastern European countries tend to have significantly higher numbers of female producers' (pg. 2). Canada is experiencing a similar rise in the number of female producers, as this demographic now makes up nearly 28% of Canadian farmers (Statistics Canada, 2018). Historical numbers in Australia were difficult to analyze, as women's farm work was not recorded in public records for many years (Alston et al., 2018). However, in a push to increase diversity in the sector, Australia recently began collecting data. They reached gender parity in agricultural education programs by 2003 and women in 2017 are 32% of the agricultural workforce (Pratley, 2017). The United States is unique, as the majority of operations are family farms. In this context, the typical farmer was historically a white male in his sixties (Hoppe and Korb, 2013). However, survey data focused on land ownership (rather than labor) and thus excluded working farmers who were not property owners. Recently, the USDA modified their surveys to capture a wider range of contributors (Glazebrook et al., 2020; USDA, 2017). In this context, data specifically on women farmers were not collected until 1978 (Hoppe and Korb, 2013). This brief overview of demographics is meant to place the subsequent discussion of gender dimensions of agricultural ethics in context. But first, we must define agricultural ethics, before exploring the birth of gender-focused research.

3 Brief overview of agricultural ethics

By the 1990s, agricultural ethics was firmly established in academia as a discipline. Research universities from Texas A&M in the US to Wageningen University in the Netherlands offer classes on the subject, though the number is dropping (see Zimdahl's chapter on institutionalizing agricultural ethics for a detailed analysis). In addition, since the publishing of Rachel Carson's Silent Spring in 1962, which included a critique of pesticide usage, public interest continues to increase. Today, articles and books on agricultural ethics are easily found in mainstream magazines and on bestseller lists. The popular work of authors, such as Paul Thompson (2015), Michael Pollan (2006), and Kingsolver (2009), helped to spark the current obsession with how food is produced. This interest is reflected in the public demand for food labeling (organic, humanely raised, etc.), as well as the growth of farmers markets and local food movements. While this change is impressive, it is important to recognize that agricultural ethics is one of the oldest philosophical topics, with discussions going back over 2000 years (Thompson and Noll, 2015). Prevalent themes in the historical discussion are central to understanding current work in agricultural ethics.

Ancient Greek thinkers, such as Xenophon (430–354 BC) and Cato the Elder (231–149 BC), wrote lengthy works on agriculture and explored the connections between food production and social systems and values (Thompson and Noll, 2015). Mirroring modern work in the field, Aristotle was one of the first scholars to discuss the connections between agriculture, daily activities, and political realities. Brief discussions of agriculture continued throughout the middle ages and into the nineteenth century. For example, John Locke (1632–1704) was deeply concerned with efficient land use and agricultural production in the

Second Treatise of Government (1689). Thomas Jefferson (1743–1826) not only farmed his Monticello plantation but also wrote extensively on agriculture. His detailed analysis of how farming is connected to democratic governance still deeply informs the literature. Similarly, the French political philosopher Baron de Montesquieu (1689–1755) argued that methods of agriculture influence national identity and social structure in the Spirit of the Laws (1748). In his political writings, Friedrich Hegel (1770–1831) built off Aristotle's work, arguing that land type and climate influence political character. According to Thompson and Noll (2015), 'this work in the Greek and European traditions of philosophy anticipates the contemporary debate over the character of rural areas and the preservation of the family farm' (p.x). Similarly, other major themes in today's agricultural literature have deep roots in historical discussions.

For instance, discussions concerning food security featured prominently in the literature of the eighteenth century (Thompson and Noll 2015) and play a key role in gender-focused research. Economists, such as Adam Smith (1723–1790) and Thomas Malthus (1766–1834), were some of the first scholars to engage with the topic of hunger. Malthus' work is very important for current discussions, as he identified the connection between agricultural technology and population growth, as the latter tends to outstrip the former resulting in famine. This work was discussed in depth by Jeremy Bentham (1748–1832) and John Stuart Mill (1806–1873), as famine and 'surplus population' were key themes in their work. Karl Marx also engaged with the twin themes of agriculture and economics in his extensive work. The discussion continued into the modern age, with Peter Singer, Amartya Sen, Martha Nussbaum, Jeffery Sachs, and other prominent scholars engaging with the topic of hunger. However, it should be noted that the discussion moves from 'carrying capacity' to questioning whether we have an ethical duty to address food insecurity, using economic and other means. Garrett Hardin (1915–2003) stresses the concern that growing populations will outstrip agricultural productivity, bringing the classic arguments of Malthus into conversation with newer work, such as Peter Singer's 'Famine, Affluence, and Morality' (1972). Singer argues that we have a duty to help those in need, irrespective of distance, while Hardin vehemently rejects this position in his work.

Work on hunger is but one topic among many in the field. The potential range of ethical issues in agriculture is extraordinary. According to Thompson and Noll (2015), these issues broadly fall within the following categories: a) issues related to food security and human health; b) issues concerning the historical, cultural, and social effects of agriculture; and c) environmental effects of food production. The work discussed above falls squarely in the first category, but this work also includes contributions by leaders of the 'organic movement', such as Liberty Hyde Bailey and Henry Wallace. Environmental effects of agriculture are extensive. Since the publishing of Rachel Carson's Silent Spring in 1962, a common theme found in popular work involves critiques of pesticides on human and wildlife populations, as inputs may have harmful effects when improperly stored, handled, or applied. In addition to the three categories, another important theme in the literature that will only be touched upon here is animal welfare. With roots in the anti-vivisectionist movement in the nineteenth century, current normative arguments extend their analysis to agricultural contexts, specifically targeting animal husbandry practices. The work of both Peter Singer (2015) and Tom Regan (1995) played an important role in the animal liberation movement, while moderate voices such as Bernard Rollin supported more humane animal agricultural practices.

4 Agricultural ethics and gender dimensions

One of the most recent themes in agricultural ethics is the growing work on gender dimensions in agriculture, which began to take shape in the late 1970s and early 1980s (Sachs, 1983). When placed in historical context, this theme is relatively new. However, it has deep roots, as it builds on the historical discussions concerning the connections between agriculture, social systems, and values. In this vein, Wendell Berry wrote extensively (and controversially) on gender and agriculture in The Unsettling of America (1978) and in other works. However, subsequent research on gender in agriculture is unique, as it embraces specific ethical commitments, grounded in the feminist movement. Ethical considerations in this literature span the gamut, but most research is guided by commitments to at least one of the following: a) gender equality, b) the desire to recognize women's contributions to agriculture and/or remove barriers to participation, and c) a demand that new agricultural technologies are not harmful to women. From a wider perspective, these commitments draw from rights-based ethical traditions - traditions, that at their core, are committed to protecting the basic rights of individuals. Throughout the world, there is a long history of using institutional power to deny specific groups basic civil rights, such as ensuring the safety, liberty, and health of citizens. Thus, one theme in feminist ethics is the commitment to protect the basic rights of women, such as the right to vote and protection from discrimination. Another theme in rights-based ethics includes concerns over risk. In addition to ethical issues regarding civil rights, Kristin Schrader-Frechette (1999) makes the argument that basic rights are violated when risks are involuntarily imposed, rather than voluntarily chosen. Similarly, the Indian scholar Vandana Shiva (2016) has been vocal in condemning the Green Revolution on grounds of social inequality and potential harms to rural communities and women, in particular. In addition to civil liberties, research on gender dimensions of agriculture is also sensitive to the effects of how risks are assessed and mitigated in society, as well as how consequences are socially distributed.

Research of this scope requires both normative commitments and empirical work, as factual evidence is imperative for determining group contributions and weighing the consequences of risk assessments. Some of the first accounts of women in agriculture in the 1980s focused on determining what contribution women made to agriculture and the identification of barriers limiting contributions. For instance, the rural sociologist Carolyn E. Sachs produced one of the first studies specifically on women in agriculture in 1983. In *Invisible Farmers: Women in Agricultural Production*, Sachs discusses the rationale for her work. Sachs describes women as 'invisible farmers', stating that 'women have always played an integral part in farming and agricultural production... Changes in agriculture are affecting the lives of millions, both men and women, but it is still men who control the land, the cash crops, the machinery, and the workers' (pg.1). Sachs was one of the first to argue that in-depth research on current and historical agricultural contexts is required to fully recognize women's contributions. The invisibility of women's labor is a driver and larger symptom of bias against women farmers and contributes to their marginalization globally (Glazebrook et al., 2020). This initial foray helped to highlight the need to understand social influences on agriculture, especially when these factors act as barriers to the recognition of women's labor. Marilyn Waring's *If Women Counted* (1988) illuminates how measuring work in terms of gross national product (or GDP) hides women's essential contributions to the economy. The world is dependent on women's work at home, in communities, and in agricultural production. But there is a legacy of women's contributions not being equally valued. Various institutionalized practices and social norms could be producing contexts where group contributions are discounted.

5 Gender and development literature

Recent research similarly focuses on uncovering injustices in resource distribution and recognition in the agricultural sector. Gender analyses have been particularly influential in global development policy and practice. As discussed above, women make up an average of 43% of the global agricultural labor force but have less access to opportunities and resources to improve productivity (FAO, 2022). The FAO estimates that closing the gender gap could increase yields by 20–30% on farms and raise total agricultural output by 2.5-4% in developing countries. These potential gains are predicated on research that suggests that women will be nearly as productive as male farmers, when provided access to similar resources (Ross, 2018). In addition, as food insecurity is linked to civil unrest, improvement can have far-reaching consequences beyond the household (Farhall and Rickards, 2021; Parpart et al., 2002; Pahl-Wostl, 2019). According to Glazebrook et al. (2020), we 'must overcome gender bias and cross barriers that block access to virtually every aspect of their work–land, finance, resources, workload, and leadership. Removing these barriers is key to obtaining gender equity in agriculture' (p.16). These statistics and recommendations are grounded in previous research in gender and agriculture. Since the 1970s, feminist scholars have been working to better understand the realities of gendered social dynamics in development practice and theory (Beneria et al., 2016) and improve interventions.

The subject of hunger features prominently in research on gender and agriculture in the global South. Thus, it builds off the long history of normative work on food security discussed above. Early work in the 1970s (such as by Peter Singer and Garrett Hardin) focused on determining whether we have a duty to help those who are suffering from a lack of basic necessities. Since the founding of the United Nations in 1945, however, the question was answered in the affirmative. By the 1975 World Conference on Women in Mexico, the key role that women's participation plays in rural development became clear (FAO, 1990). By 1990, women's role in agricultural development was firmly established in the policy literature. For example, the report titled Women in Agricultural Development, Gender Issues in Rural Food Security in Developing Countries states the following:

> [D]evelopment interventions to reduce women's drudgery can significantly enhance their contribution to household food security. The provision of water supplies; the introduction of light transport facilities for the carrying of firewood, farm produce and other loads; the introduction of labour saving agricultural tools, the introduction of grinding mills and other crop processing equipment are crucial means of freeing Women's time. Such technologies not only create possibilities for women to enter into more income-generating activities, but also help in reducing their stress and in improving the health and nutrition of women and children. (FAO, 1990, N.P.)

The recognition of connections between women's labor and food security is an important part of the literature. Even today, in the global South, the 'gender gap' remains in agriculture as women still face inequalities that effect their access to resources and food security (Glazebrook, 2011; Johnson-Welch et al., 2000). A 2017 survey of 104 countries found that women own less than 13% of agricultural land, while they constitute 1.4 billion of the 2.5 billion people who rely on agriculture for subsistence (Glazebrook, 2011; Balakrishnan and Fairbairn-Dunlop, 2005). The survey also found that women are more susceptible to experience insecurity, are less able to obtain resources, and have reduced agency in household and community decision-making. Additional research supports these conclusions, as it has been argued that women in Latin America, Africa, and Asia have less access to agricultural research, technology, livestock, water, credit, and land, among other key resources (Balakrishnan and Fairbairn-Dunlop, 2005; Vaque, 2017). While women often shoulder a greater responsibility for household food security, they are typically under-resourced and marginalized.

The desire to ensure that communities have access to basic needs guides much of this work. However, further research is concerned that targeting women in development may treat them as merely instruments or a means to an end. For example, in *Women and Human Development: The Capabilities Approach*, Martha Nussbaum argues that it is imperative for international development circles to approach gender difference as a problem of justice (Nussbaum, 2000). Economists have long relegated justice-related arguments to the background in policy discussions. However, this is highly problematic when providing gender-focused recommendations, as unequal social, economic, and political circumstances indelibly mark the lives of women around the world. Women are often treated like second-class citizens in many countries, leading Nussbaum to argue that these external circumstances effect the development of key capabilities needed to live a full life. Thus, having access to the full range of opportunities and liberties would not only improve external circumstances but also positively effect their inner lives. In short, it will empower women by fundamentally enriching what they believe they can achieve and what they dare to hope for. Recognizing the connection between women's work and food security is a good first step. However, it is problematic if women are not treated as an end in their own right but merely as a means to ensuring food security. Instead, development programs should work toward empowering people to live complete lives, with fully developed capabilities.

This turn toward empowerment was foreshadowed in Vandana Shiva's 1988 publication *Staying Alive: Women, Ecology, and Development*. In this publication, development initiatives are described as post-colonial projects forced on the global South in an attempt to force Western conceptions of progress, indelibly tied to colonialism. There's a lot to unpack here, but respecting the choices and autonomy of rural communities is a prominent thread through this work. Today, this focus on empowerment is indelibly connected to the trifecta of gender, agriculture, and development. Empowerment is now a mainstay on the international development agenda, as goal five of the United Nations 'Sustainable Development Goals' intimates (UN Economic and Social Council, 2017). Specifically, it pushes global communities to prioritize gender interventions, in order to 'achieve gender equality and empower all women and girls' (UN, 2015). According to Farhall and Rickards (2021), 'efforts to implement the gender agenda into agriculture has, at least in terms of visibility, been a resounding success' (p.2). Important research areas include microcredit as a tool for empowerment (Bahta et al., 2017), gender gaps in agricultural knowledge and training (Zossou et al., 2017), small-holder practices that affect gender equity (Farnworth et al., 2018), gendered dynamics of climate change adaptation and sustainability (Jerneck, 2018), among others. When taken as a whole, it is clear that gender plays a key role in agricultural development initiatives, beyond merely increasing access to agricultural resources. This is not surprising, as challenges faced by female farmers in the global South increases their vulnerability to fluctuating weather patterns, undermines their resilience and autonomy, and harms food security in their communities.

Research on gender and development continues to advance, as scholars critique previous policies and interventions. For instance, feminist critiques of current programs argue that international organizations may be reproducing neoliberal policy frameworks, rather than taking meaningful steps toward gender parity in sustainable agriculture (Farhall and Rickards, 2021). Thus, an important theme in current research is critically analyzing the intellectual underpinnings of 'gender-focused' programs (Cornwall and Rivas, 2015), especially interventions that utilize dominant paradigms of food, policy, and agriculture. This work is grounded in earlier arguments that development programs are often unsuccessful, if they inadequately engage with gender, as embedded in wider development discourse. As Farhall and Rickards (2021) argue, 'un-nuanced gendered approaches to development can exacerbate inequalities, re-entrench forms of difference, or marginalize women in new ways, the assumptions and outcomes of gender initiatives require constant evaluation' (p.3). Simply adding women is inadequate to address gender injustices (Moser, 1989; Parpart et al., 2002). Thus, vigilance is imperative for any meaningful transformation of food systems, from a gendered perspective. Research at the intersections of gender, agriculture, and development is on-going. While great strides have been made globally, a 'gender gap' remains in agriculture, as women still face a myriad of inequities (Johnson-Welch et al., 2000).

6 Gender dimensions in the global North

Similarly, great strides have been made in research on gendered dimensions of agriculture in the global North. According to recent census data, countries in this context are experiencing a resurgence of female farmers, as the EU, the United States, Australia, and many other countries are working toward achieving gender parity in agriculture (Glazebrook et al., 2020). In fact, recent survey data clearly show an increase in diverse practitioners as more women take leadership positions in farming operations. For example, in Canada, women make up approximately 28% of farmers (Statistics Canada, 2018). Similarly, the United States is seeing an increase in female producers, with women making up 36% of the total workforce in 2017 (USDA, 2017). This increase is impressive, as farms in the global North are predominantly large-scale commodity operations, in contrast to the prevalence of subsistence agriculture in the global South. The increase of diverse farmers can be partially contributed to a) an increase in recognition of women's contributions to agriculture and b) the identification or removal of barriers inhibiting entry into the field (Glazebrook et al., 2020). Thus, one can argue that early research on gender in agriculture in the late 1970s and 1980s paved the way for the 'feminization of agriculture' in the global North.

In particular, Waring's (1988) work on how economic frameworks actively hide women's contributions foreshadowed later critiques of agricultural data collection methodologies. On the ground, this type of research resulted in a more nuanced portrait of the agricultural sector. For instance, in the United States, gender-sensitive data were not collected until 1978 (Hoppe and Korb, 2013). In addition, historical data collection methods typically focused on land ownership, rather than contributions of labor on the farm. As the majority of agricultural operations in Canada and the United States are classified as family farms (USDA, 2017; Fremstad and Paul, 2020), this meant that considerable contributions of family members (not deemed the owner) and other workers were excluded. In several instances, the owner might not live on the farm. In 2017, the USDA modified their surveys to provide a more nuanced understanding of agricultural labor (Hoppe and Korb, 2013). Specifically, the USDA allowed survey takers to report multiple primary producers and managers, rather than only one primary producer for the farm (USDA, 2019). This change resulted in an increase in the number of women reported, which is not surprising as the division of labor in families includes female participation. Thus, the sharp increase in female producers could be the result of the refinement of surveys and other data collection tools.

The improvement of data collection methods is important but only one area of research on gender and agriculture in the global North. Various social practices have been subjected to critical analysis, with an emphasis on gender and inclusion. To return to the example of labor contributions on the farm, farming is traditionally considered a male-dominated profession in the global North, largely due to land transfer practices and historical patriarchal laws (Fremstad and Paul, 2020; Ramey, 2014; Whatmore, 1991). While farming is a family affair in this context, there is a legacy of women's contributions being undervalued. Literature on social realities of farming highlight inequalities concerning how contributions are recognized, as men were often seen as 'farmers', while women were typically conceptualized as 'farmwives' in dominant discourse (Shortall, 1992). As Fremstad and Paul (2020) argue, 'despite the fact that the family farm depended on women's work for survival, women continued to work as invisible farmers' (p.6). Biases and gender norms resulted in women's contributions being ignored in national farm policy, agricultural knowledge exchange, and extension programs (Leckie, 1996). Research documenting this erasure prompted the USDA to label women socially disadvantaged farmers and thus make targeted loan funds available to offset barriers (USDA, 2016).

Despite steady gains in the recognition of women's work, recent literature on barriers highlights several areas of concern. For example, there is a growing body of work exploring how gendered conventions effect formal and informal practices (Carter, 2019; Drunschke and Secchi, 2014; Twyman et al., 2015). Carter (2019) argues that contextual factors constrain women's management options in several key ways. First, women were historically excluded from spaces that play a key role in the dissemination of resources, such as research stations, extension offices, and other spaces where agricultural knowledge is produced and exchanged (Leckie, 1996; Sachs, 1983). The result of this exclusion was a hyper-focus on men's needs and experiences, even in agricultural sectors where women make up the majority of producers, such as in sustainable and organic agriculture (Peter et al., 2000; Carolan, 2005). The prioritization of large-scale agriculture over other types of agriculture prioritized by female producers (Fremstad and Paul, 2020; Carter, 2019) is linked to a reduction of management options available. In this vein, women report that it is common for agency staffers and male farmers to pressure them into adopting management strategies not aligned with their goals. These social reactions reinforce the prioritization of men's farming needs, as increasing crop yields is stressed by this demographic, in contrast to other priorities stressed by women, such as community and environmental health and food security (Chiappe and Flora, 1998). Gender-based exclusion and problematic social reactions are also linked to perceived illegitimacy of women as leaders in agricultural contexts (Alsgaard, 2012; Carter, 2019) and to a reduction of access to farm loans and other key resources (USDA, 2022). For instance, the USDA has a history of being accused of denying loans to female farmers at a higher rate than male farmers (USDA, 2016). In short, the social processes of boundary maintenance and maintaining gendered agricultural landscapes is still common in the global North (Carter, 2019). Thus, research on gender bias in agriculture in the global North is important for addressing systemic and institutionalized discrimination.

These formal and informal practices are clearly linked to disparities in farming income. According to Fremstad and Paul (2020), 'farms operated by women earn forty percent less farm income than farms operated by men after controlling for farm and operator characteristics. These findings indicate that farming is one of the most unequal professions in the United States today' (pg. 124). Key factors that influence pay disparities include a) the fact that women are more likely to operate sustainable agriculture or community-focused farming endeavors and b) women's lack of access to land, thus forcing them to place fewer acres under cultivation. The links between sustainable agriculture and farm income may not be apparent for some readers. Sustainable agriculture practices include using organic methods, running Community Supported Agriculture operations, and selling direct-to-consumer. These practices are often guided by the desire to support community health, reduce negative environmental effects of production, improve regional resilience, etc. (Delind, 2011). However, sustainable agriculture methods are also associated with lower overall profits, which in turn reduce their support by institutions (Fremstad and Paul, 2020).

7 Conclusion and future trends

Research on gender dimensions in agriculture is on-going. With this being said, great strides have been made since research began to take shape in the late 1970s and early 1980s. It is difficult to convey the importance of initial studies, such as those by Carolyn E. Sachs, Marilyn Waring, Vandana Shiva, and others. This and later work helped academics and policymakers develop a more nuanced portrait of the agricultural sector and develop equity-focused policies and programs. It also helped to make previously 'invisible' contributions visible, by identifying social and institutional practices that conceal women's contributions. In the global context, development programs are now highly sensitive to the trifecta of gender, agriculture, and development. Gender empowerment is now a mainstay on the international development agenda. Gender-sensitive programs are often guided by the intertwined goals of empowering women and increasing food security in communities. Thus, research on gender dimensions in agriculture has been highly influential, guiding programs and policies that directly effect individual lives.

Changes in the agricultural landscapes of countries in the global South and North are impressive. However, future directions in research could include further work on social and institutional barriers that are harmful to specific demographics. Additionally, cross-comparisons of strategies employed in different contexts may be useful, as policymakers and practitioners could learn from local failures and successes. Women around the world share access and inclusion challenges. However, they often do so for very different reasons. For instance, women often choose to farm organically in the global North. In contrast, women in the global South often practice these methods largely by default, if they do not have access to necessary inputs (Glazebrook et al., 2020). Connectedly, research improving sustainable agriculture methods could be helpful, as a higher percentage of diverse practitioners contribute to this type of production (Chiappe and Flora, 1998). Women's approaches to and experiences in agriculture are different based on location, socio-political factors, and training. Better understanding how local differences influence the daily lives of producers is necessary to advance our understanding of gender dimensions in agriculture.

8 Where to look for further information

- Gender, development and globalization: Economics as if all people mattered, by Beneria, et al. (2016).
- Contemporary feminist analysis of Australian farm women in the context of climate changes, by Alston, Clark, and Whittenbury (2018).

- Gender and social norms in Agriculture: A review, by Hillenbrand and Miruka (2019).
- Relational agriculture: Gender, sexuality, and sustainability in U.S. farming, by Leslie, Wypler and Bell (2019).
- Closing the gender gap in agriculture under climate change, Nyasimi and Huyer (2017).
- Agricultural News from Latin America and the Caribbean. Rural Women– A Key Asset for Growth in Latin America and the Caribbean, by Agronoticias. Available online: http://www.fao.org/in-action/agronoticias/detail/en/c/501669/.
- Food Sovereignty and gender justice, by Portman (2018).
- Gender and the American Family Farm in the 20th Century, by Ramey (2014).
- Learn the Facts: Rural Women and Girls, United Nations (2018). Available online: https://www.unwomen.org/en/digital-library/multimedia/2018/2/infographic-rural-women.
- Women Facts and Figures: Economic Empowerment, United Nations Women (2018). Available online: https://www.unwomen.org/en/what-we-do/economic-empowerment/facts-and-figures.

8.1 Key associations and journals

- IDEA (International Development Ethics Association), attended by scholars focused on the global South.
- EurSafe (European Society for Agriculture and Food Ethics), attended by scholars focused on a wide range of gender dimensions.
- AFHVS (The Agriculture, Food, and Human Values Society), attended by scholars focused on food in the U.S. context.
- ASFS (Association for the Study of Food and Society) promotes the interdisciplinary study of food and society.
- *Journal of Agricultural and Environmental Ethics* provides a venue to publish work on environmental ethics, including research on gender dimensions.
- The *Journal of Agriculture and Human Values* provides a venue to research on gender and society.
- Culinary Mind: The Center for Philosophy of Food provides a platform to philosophically think about food systems.
- Signs: Journal of Women. In Culture and Society publishes work on gender and development.
- Hypatia: A Journal of Feminist Philosophy publishes work with a gendered component.

9 References

Alsgaard, H. (2012). Rural inheritance: Gender disparities in farm transmission. *North Dakota Law Review* 88, 347–408.

Bahta, Y. T., Strydom, D. B. and Donkor, E. (2017). "Microcredit and gender empowerment: Policy implications for sustainable agricultural development" in *Eritrea. Development in Practice* 27(1), 90–102. https://doi.org/10.1080/09614524.2017.1259393.

Balakrishnan, N. and Fairbairn-Dunlop, P. (2005). *FAO Regional Office for Asia and the Pacific: Prospects and Paradoxes*. Food and Agriculture Organization, Regional Office for Asia and the Pacific, Bangkok, Thailand.

Beneria, L., Berik, G. and Floro, M. S. (2016). *Gender, Development and Globalization: Economics as If All People Mattered*. Routledge, New York, NY.

Berry, W. (1978). *The Unsettling of America: Culture & Agriculture*. Sierra Club Books.

Carolan, M. S. (2005). Barriers to the adoption of sustainable agriculture on rented land: An examination of contesting social fields. *Rural Sociology* 70(3), 387–413.

Carson, R. (1962). *Silent Spring*. Houghton Mifflin Harcourt Trade & Reference Publishers.

Carter, A. (2019). We don't equal even just one man: Gender and social control in conservation adoption. *Society of Natural Resources* 893-910.

Chiappe, M. B. and Flora, C. (1998). Gendered elements of the alternative agriculture paradigm. *Rural Sociology* 63(3), 372–393.

Cornwall, A. and Rivas, A. M. (2015). From 'gender equality and 'women's empowerment' to global justice: Reclaiming a transformative agenda for gender and development. *Third World Quarterly* 36(2), 396–415.

DeLind, L. B. (2011). Are local food and the local food movement taking us where we want to go? Or are we hitching our wagons to the wrong stars? *Agriculture and Human Values* 28(2), 273–283.

Drunschke, C. G. and Secchi, S. (2014). The impact of gender on agricultural conservation knowledge and attitudes in an Iowa watershed. *Journal of Soil and Water Conservation* 69(2), 95–106.

European Commission (2019). Females in the field: More women managing farms across Europe. Available at: https://ec.europa.eu/info/news/queens-frontage-women-farming-2019-mar-08_en (accessed 27 May 2020). European Commission.

Farhall, K. and Rickards, L. (2021). The "Gender agenda" in agriculture for development and its (lack of) alignment with feminist scholarship. *Frontiers in Sustainable Food Systems* 5. https://doi.org/DOI=10.3389/fsufs.2021.573424.

Farnworth, C. R., Stirling, C. M., Chinyophiro, A., Namakhoma, A. and Morahan, R. (2018). Exploring the potential of household methodologies to strengthen gender equality and improve smallholder livelihoods: Research in Malawi in maize-based systems. *Journal of Arid Environments* 149, 53–61.

Food and Agriculture Organization of the United Nations (1990). Women and sustainable food security. Available at: https://www.fao.org/3/x0171e/x0171e02.htm (accessed 5 January 2022).

Food and Agriculture Organization of the United Nations (2011). *Women in Agriculture Closing the Gender Gap for Development.* Food and Agricultural Organization of the United Nations, Rome, Italy.

Food and Agriculture Organization of the United Nations (2012). *Gender Inequalities in Rural Employment in Ghana: An Overview.* Gender, Equity and Rural Employment, Rome, Italy.

Food and Agriculture Organization of the United Nations (2022). Women in agriculture. Available at: https://www.usda.gov/media/press-releases/2019/04/11/2017-census-agriculture-data-now-available (accessed 5 January 2022).

Fremstad, A. and Paul, M. (2020). Opening the farm gate to women? The gender gap in U.S. agriculture. *Journal of Economic Issues* 54(1), 124–141.

Glazebrook, T. (2011). Women and climate change: A case study from Northeast Ghana. *Hypatia* 26(4), 762–782.

Glazebrook, T., Noll, S. and Opoku, E. (2020). Gender matters: Climate change, gender bias, and women's farming in the global South and North. *Agriculture* 10(7), 1–25.

Government of Canada (2022). Overview of Canada's agriculture and agri-food sector Available at: https://agriculture.canada.ca/en/agriculture-and-agri-food-canada/canadas-agriculture-sectors/overview-canadas-agriculture-and-agri-food-sector (accessed 27 May 2022).

Hoppe, R. and Korb, P. (2013). *Characteristics of Women Farm Operators and Their Farms* (Economic Information Bulletin No. EIB-111). United States Department of Agriculture, Washington DC.

Jerneck, A. (2018). Taking gender seriously in climate change adaptation and sustainability science research: Views from feminist debates and sub-Saharan small-scale agriculture. *Sustainability Science* 13(2), 403–416. https://doi.org/10.1007/s11625-017-0464-y.

Johnson-Welch, C., Alemu, B., Msaki, T. P., Sengendo, M., Kigutha, H. and Wolff, A. (2000). Improving household food security: Institutions, gender, and integrated approaches. http://www.hubrural.org.*Imaging*/pdf/basissem9911_panel3c.pdf (accessed 27 May 2020).

Kingsolver, B., Kingsolver, C. and Hopp, S. L. (2009). *Animal, Vegetable, Miracle: A Year of Food Life*. Harper Collins.

Leckie, G. (1996). They never trusted me to drive: Farm girls and the gender relations of agricultural information transfer. *Gender, Place, and Culture: A Journal of Feminist Geography* 3, 309–326.

McAfee, N. (2018). Feminist philosophy. In: Zalta, E. N. (Ed.), *The Stanford Encyclopedia of Philosophy*. Stanford University, Stanford.

Moser, C. O. N. (1989). Gender planning in the third world: Meeting practical and strategic gender needs. *World Development* 17(11), 1799–1825. https://doi.org/10.1016/0305-750X(89)90201-5.

Nussbaum, M. (2000). *Women and Human Development: The Capabilities Approach*. Cambridge University Press, New York.

Nyasimi, M. and Huyer, S. (2017). Closing the gender gap in agriculture under climate change. *Agricultural Development* 37–40.

Pahl-Wostl, C. (2019). Governance of the water-energy-food security nexus: A multi-level coordination challenge. *Environmental Science and Policy* 92, 356–367.

Parpart, J. L., Rai, S. M. and Staudt, K. (2002). *Rethinking Empowerment: Gender Development in a Global/Local World*. Routledge, New York, NY.

Peter, G., Bell, M. M., Jarnagin, S. and Bauer, D. (2000). Coming back across the fence: Masculinity and the transition to sustainable agriculture. *Rural Sociology* 65(2), 215–233.

Pollan, M. (2006). *The Omnivore's Dilemma: A Natural History of Four Meals*. Penguin.

Pratley, J. E. (2017). Agriculture – From macho to gender balance. In: *Proceedings of the 18th Australian Society of Agronomy Conference* (Vol. 4).

Ramey, E. (2014). *Class, Gender and the American Family Farm in the 20th Century*. Routledge, London.
Regan, T. (1995). Obligations to animals are based on rights. *Journal of Agricultural and Environmental Ethics* 8(2), 171–180.
Ross, C. R. (2018). Women and agricultural productivity: Reframing the issues. *Development Policy Review* 36(1), 35–50.
Sachs, C. (1983). *The Invisible Farmers: Women in Agricultural Production*. Rowman & Allanheld, Totowa, NJ.
Sassenrath, G. F., Heilman, P., Luschei, E., Bennett, G. L., Fitzgerald, G., Klesius, P., Tracy, W., Williford, J. R. and Zimba, P. V. (2008). *Technology, Complexity and Change in Agricultural Production Systems*. Cambridge University Press, Cambridge.
Schrader-Frechette, K. (1999). *Risk and Rationality: Philosophical Foundations for Populist Reforms*. University of California Press, Berkeley.
Shiva, V. (2016). *The Violence of the Green Revolution: Third World Agriculture, Ecology, and Politics*. University Press of Kentucky, Kentucky.
Shortall, S. (1992). Power analysis and farm wives. *Sociologia Ruralis* 32(4), 431–451.
Singer, P. (1972). Famine, affluence, and morality. *Philosophy and Public Affairs* 1, 229–243.
Singer, P. (2015). Animal liberation: The definitive classic of the animal movement. *Open Road Media*.
Statistics Canada (2018). Gender, diversity and inclusion statistics. Available at: https://www.statcan.gc.ca/eng/topics-start/gender_diversity_and_inclusion (accessed 27 May 2020).
Thompson, P. B. (2015). *From Field to Fork: Food Ethics for Everyone*. Oxford University Press, Oxford.
Thompson, P. B. and Noll, S. (2015). Agriculture ethics. In: *Ethics, Science, Technology, and Engineering: An International Resource*. Cengage Press, Independence, pp. 35–42.
Twyman, J., Muriel, J. and Garcia, M. A. (2015). Identifying women farmers: Informal gender norms as institutional barriers to recognizing women's contributions to agriculture. *Journal of Gender Agriculture and Food Security*. https://doi.org/10.22004/ag.econ.246047.
UN Economic and Social Council (2017). *Progress Towards the Sustainable Development Goals (No. E/2017/66)*. United Nations, New York, NY.
United Nations (2015). *Sustainable Development Goals: Transforming Our World: The 2030 Agenda for Sustainable Development* (No. A/Res/70/1). United Nations, New York, NY.
United Nations (2019). Women's progress on the sustainable development goals: The gender snapshot 2019. Available at: https://www.unwomen.org/en/digital-library/publications/2019/09/progress-on-the-sustainable-development-goals-the-gender-snapshot-2019 (accessed 5 January 2022).
United States Department of Agriculture (2016). *Hispanic and Women Farmers and Ranchers Claim Resolution Process (No. 50601- 0002-21)*. United States Department of Agriculture, Washington, DC.
United States Department of Agriculture (2019). 2017 census of agriculture data. Available at: https://www.usda.gov/media/press-releases/2019/04/11/2017-census-agriculture-data-now-available (accessed 27 May 2020).
Vandana, S. (1988). *Staying Alive: Women, Ecology, and Development*. North Atlantic Books, New York.

Vaque, J. A. (2017). Rural women—A key asset for growth in Latin America and the Caribbean. *Agricultural FAO News Form Latin America and the Caribbean*.

Waring, M. (1988). *If Women Counted: A New Feminist Economics*. Harper & Row, New York, NY.

Whatmore, S. (1991). *Farming Women: Gender, Work, and Family Enterprise*. Macmillan, London.

World Bank (2014). *For Up to 800 Million Rural Poor, a Strong World Bank Commitment to Agriculture*. The World Bank. Available at: https://www.worldbank.org/en/news/feature/2014/11/12/for-up-to-800-million-rural-poor-a-strong-world-bank-commitment-to-agriculture (accessed 1 May 2022).

Zossou, E., Arouna, A., Diagne, A. and Agboh-Noameshie, R. A. (2017). Gender gap in acquisition and practice of agricultural knowledge: Case study of rice farming in West Africa. *Experimental Agriculture*: 566–577. https://doi.org/10.1017/S0014479716000582.

Chapter 5

Agricultural ethics – the farmer's perspective

Robert L. Zimdahl, Colorado State University, USA

1 Introduction

Agriculture's ethical issues are as old as farming, but in the last 50 years, there has been only limited interest in agricultural ethics. Agriculture, the essential human activity, is the most widespread human environmental interaction. Those engaged in agriculture continue to improve and justify production practices. Analysis of agricultural practices and critiques leads one to conclude there are important, often unaddressed, ethical problems. Production alone does not justify every practice. There is a need for an explicit ethical foundation to debate and rank the priority of production, protection of human health, the environment, and the need of all humans for food.

The noun 'ethics' is used to support what the advocates and proponents of some technology or practice think is the right thing to do. Ethics is used to justify their advocacy because it is assumed everyone knows if it is ethical it must be good and therefore right (Thompson, 2020). Those who disagree about the use of some technology or advocate different technology use ethics or ethical to justify their position. They may not know how or they do not think it is necessary to proceed to any moral argument about why what they advocate

http://dx.doi.org/10.19103/AS.2023.0125.05

is ethically correct (promotes happiness over suffering, should be universally accepted, is virtuous, etc.). Use of the word is sufficient justification.

The practices farmers employ to produce food and fiber have economic, political, social, and moral dimensions (Andrée, 2021). Durning (1994) highlighted how profound these ethical dilemmas might be in forcing a fundamental reappraisal of agriculture and the modern consumer society that many have and many aspire to obtain.

> Limiting the consumer lifestyle to those who already have it is not politically possible, ecologically efficient or ethically defensible. Extending that lifestyle to all would hasten the demise of the ecosystem we all depend on. It would hasten the ruin of the biosphere. On the other hand, reducing the consumption level of the consumer society, and tempering material aspirations elsewhere, though morally acceptable, is a quixotic proposal. It bucks the trend of centuries, but it may be the only option. (Durning, 1994)

Some of the issues to ensure agriculture's future development is ecologically sustainable and socially equitable are addressed by the United Nations Sustainable Development Goals (SDGs) (https://www.un.org/sustainabledevelopment/) (2017). Moyer and Hedana (2020) highlighted the major changes in domestic and international policy required to implement the sustainability agenda. There are different visions of agriculture to address these challenges, ranging from a traditional 'productivist' paradigm where the primary justification is the continued availability of nutritious cheap food for all. The opposing view is that production must be balanced by ecological considerations with a focus on protection and stewardship of natural resources (see Leopold, 1966). Others suggest future agriculture needs to return to small-scale farming serving local markets and a re-engineering agriculture and other industries to create a circular bio-based economy focused on use, recycling, and re-use of organic raw materials (Bowens, 2015; Grover and Gruver 2017; Wensing et al., 2019).

Gremmen (2020) argues farmers have a wealth of experience and skill, yet they may not do the right things because of complex choices they face. All choices have an ethical dimension, but in Gremmen's view, all lack a moral compass to guide them toward achieving sustainable, resilient agricultural production systems. Indeed, Thompson (2020) has argued that differing players across the agri-food supply chain have claimed their practices are ethical because they meet the morally acceptable goal of feeding a growing population. However, all involved in agriculture lack an explicit ethical foundation as a guide toward agriculture's worthy goals: achieving necessary production to ensure an adequate supply of safe, nutritious food for all, protecting human health, protecting the environment and other creatures, small farms, and rural communities threatened by some of the practices modern, energy, chemical, and capital intensive farmers rely on to optimize production.

2 Understanding farmers' constraints and motivations

Policymakers and those who create and attempt to achieve allegiance to new technology 'tend to have a misleading, instrumental view of the role of farmers', as technical providers of biomass/energy feedstocks (Wensing et al., 2019; Rossi and Hinrichs, 2011). Listening to farmers and learning from them will help agriculture's policymakers achieve a clearer understanding of why farmers farm as they do. A reality of modern agriculture is most farmers do not produce food – they grow commodities which are sold to multinational corporations for conversion to what we buy in the supermarket. The narrow instrumental view ignores the role of farms as 'providers of quality food, managers of the agricultural ecosystem and landscape, and contributors to rural development' (Schmid et al., 2012). Within the agricultural sciences, there is little awareness or concern about the factors that influence farmers' decisions to resist or adopt new practices.

The author's conversations with farmers show they have rational reasons for not adopting what others regard as best management practices (e.g. crop rotation and no-till farming). My conversations suggest expensive equipment and unfamiliar ways of working are significant barriers. However, there are often broader reasons for avoiding change: their income is adequate, problems related to soil health are manageable, they are comfortable with the results of the practices they use, and are not convinced new ways are better.

Broader assumptions suggest helping farmers change their practices requires an understanding of the cultural context in which they operate. The traditional instrumental view of farmers and farming does not include an adequate understanding of the experiential and cultural factors motivating farmers to adopt or not adopt management practices that may contribute to a more sustainable future (Prokopy et al., 2008). There is a growing body of research on agricultural practices which may guide policymakers, scientists, and agribusinesses to assist farmers to change from conventional to more sustainable practices.

As with other social groups, farmers are heterogeneous with widely differing attitudes to farming and rural life. Studies of conventional and 'alternative' (e.g. organic) farmers show that the two groups have very different beliefs, values, and priorities (Petrzelka et al., 1996; Fairweather and Campbell, 2003). It is important about recognize that many farmers have an unspoken, deeply held sense of what good farming is (Lavoie and Wardropper, 2021). This image and the assumptions which underpin strategies help farmers accept and introduce change.

Coon et al. (2020) studied landowner decisions regarding grass management for cattle grazing on the eastern US Great Plains regarding decisions to control invasive grasses. They identified factors such as understanding the importance of non-native grass management, perceived ability to successfully implement non-native grass management programs, and the attitudes of neighboring ranchers. They found neighbor's expectations had more influence on action than natural resource agency recommendations.

Among the multiple constraints to adoption of more sustainable practices, Perry and Davenport (2020) identified an individual's belief in her or his capacity or ability to make changes and mistrust of government agencies which was related to neglect or ignorance by policymakers and others of farmers' goals and values. Grover and Gruver (2017) and Nleya and Clay (2021) identified a range of barriers to adoption of sustainable farm management: markets, regulations, cost and availability of equipment and labor, understanding of and interest in environmental or ecological issues, and access to education or research support. A meta-analysis of qualitative studies investigating reasons for adoption or non-adoption of conservation practices and programs (CPPs) by Ranjan et al. (2019) identified degrees of environmental awareness and trust in information sources. Barriers included familiarity with current farm management practices, negative perceptions of CPP, perception of the financial risk in implementing CPP, and land tenure arrangements. Prokopy et al. (2019) identified factors explaining farmers' motivations to adopt regenerative agricultural practices (no-till, cover crops, and crop rotation). These include self-identification as stewards of the land, non-financially motivated environmental attitudes, actively seeking more information, willingness to investigate available programs, higher income, and a higher level of formal education. Ehsan et al. (2021) found comparable differences in attitudes and practices among Pakistan farmers toward adoption of land-use practices for pasture production. Older farmers with more education, larger herds, and more milking animals were much more likely to use improved pasture practices than those with large families, high economic burdens, less satisfaction from livestock farming, and a lack of land ownership.

Understanding the ethical responsibilities and choices facing farmers includes knowing the broader constraints within which they must work and their role in the wider agri-food supply chain. It is essential to understand both long- and short-term trends in the structure and economics of farming. Farms are businesses which need to make a commercial return if they are to remain viable (Thomas and Kevan, 1993). The further one delves into the social science literature on agriculture the more one learns about the multiple constraints to adoption of conservation of natural resource practices and progress toward sustainability.

Presentation of conversion programs should shift emphasis from negative production outcomes to assured positive outcomes in efficiency and profit. Farmers want and deserve a clear discussion of risk and reward (Mayberry et al. 2005). However, there is poor understanding of the values that drive social and agricultural change and of how and why farmers decide what they must and what they can do.

3 Environmental and social costs

Agriculture's production practices have inevitable environmental (pollution from pesticides, fertilizer, animal waste, and species loss) and social costs (harm to farm labor, loss of farmers, and conversion of farmland (e.g. development)). There is a paucity of research designed to determine if the public will continue to tolerate these costs. Conversations with agricultural producers and others engaged in agriculture are not occurring. With few exceptions (*Journal of Agricultural and Environmental Ethics*), conversations and thoughtful articles have not included debate about agriculture's ethical problems. Academics, research scientists, those in supporting agricultural industries, and members of the general public have opinions about what farmers do and ought to do. There is an incomplete understanding of what motivates farmers not to adopt best management practices that lead to resolution of technological and ethical dilemmas (Prokopy et al. 2008) regarding conservation, sustainable food production, and environmental rights.

Agricultural organizations, including land-grant universities, should formulate and prominently include in their mission and objectives a statement of their ethical position relevant to the conduct of agriculture 'in a manner that makes a decent life for humans possible while, at the same time, retaining the ecological dynamics to sustain life on the planet' and make farming possible (Kirschenmann, 2000). To achieve this requires those involved at all levels of agriculture cooperate among themselves, with society as a whole and, of most importance, with a wide range of farmers and ranchers. Research and other organizations involved in agriculture should not assume a right to tell farmers and ranchers what they ought to do. Elegant, well-researched, solutions presented by agricultural scientists have often proved to be incompatible with the reality of farming and ranching (Campbell and Fairweather, 1998). Farmers and ranchers work hard to produce vital products, with most behaving in reasonable ecological, economic, and environmental ways. Some practices will have to change. But those who do research and develop technology must listen to the producers, learn from them, and help them adapt to change. Part of learning and helping is to determine whose responsibility it is to change, who pays, and who benefits?

An example of listening is the Mali government's decision not to introduce genetically modified (GM) cotton (Pimbert and Barry, 2021). Malian farmers gathered to deliberate on GM cotton and the future of farming in a group unique in West Africa and perhaps in the world - Citizens Space for Democratic Deliberation (Espace Citoyen d'Interpellation Démocratique). Men and women farmers were given an opportunity to share knowledge on the benefits and risks of GM cotton and make future policy recommendations.

4 Agricultural issues

A major ethical issue for agriculture is genetic modification of crops. GM crops have been the most rapidly adopted agricultural technology ever developed. In 2003, 67.7 million hectares of GM crops were grown in the world. By 2019, more than 190 million hectares were grown. It is projected 200 million hectares were grown in 28 nations by 2022. GM crops are accepted by farmers and regulators in 29 countries. They are not as widely accepted by consumers, many of whom have strong ethical objections to genetic modification of anything. Losses during harvest, post-harvest storage, and processing should be reduced. Agriculture could lead the way toward a true recycling economy, where the waste from one enterprise becomes the feedstock for another. The figure varies but approximately one-third of food produced for human consumption is lost due to insects, rodents, or diseases or is wasted because we do not consume all available. It is a reasonable moral claim that the poor could be but are not fed with what we waste.

There are clear, well-reasoned arguments about the necessity and moral status of animals in Confined Animal Feeding Operations (CAFOs) and equally well-reasoned arguments that animals are essential to a truly integrated, sustainable, resilient agriculture. Reduction of excessive, harmful consumption of meat in developed countries is a human health and environmental issue that is of direct concern to the public and all engaged in animal agriculture. Pesticides, especially prophylactic use, and excessive fertilizer use will have to be diminished. Erosion of soil must decrease to protect our most valuable environmental resource and diminish water and air pollution. If farmers are to conserve energy, reduce pollution, and promote ecological stability, policies rewarding them for adopting such practices will have to be developed.

My conversations with farmers reveal they know how to farm, prevent soil erosion, optimize water use, reduce pollution, and achieve animal welfare and ecological harmony. Government and market policies rewarding production must be changed to compensate farmers for adopting environmentally favorable practices. Farmers want to participate in determining how their land is managed. We should abandon the antiquated view of farmers and farms as quaint places populated by people who do not understand modern life and its values. The practices farmers employ to produce food and fiber have enduring environmental effects, raise moral questions, and have important political implications (Andrée, 2021). The land, an essential resource, cannot and should not be 'contained and controlled in ways imagined by agronomists'. We need to think carefully about how agricultural educators, researchers, and industry representatives engage with farmers (Benefeldt et al., 2021). The relationships and the advice offered must consider the farmers' experience, values, the complexity of their business, and existing social-economic relationships. This has not always occurred in practice. In the United States, it has been suggested that 'Federal agricultural policy has failed to effectively promote and reward diversification and conservation practices' (Spangler et al., 2020).

US food and fiber production is heavily concentrated in some areas on some crops. Large farms are getting larger and small farms are disappearing. This has resulted in a highly simplified monocultural agriculture dependent on chemicals, capital, and petroleum energy. Crop diversity is declining as are ecosystem services to and from agriculture. The justification has been the necessity of feeding a growing world population and satisfying market demand.

Production technology changes are illustrated by data on agrochemical use in Canada (Malaj et al., 2020). Fertilizer is applied on 78% of the total cropland. Herbicides are applied on 30% of the Western Pacific province farms and 81% of the Prairie province farms, which comprise 90% of Canada's total cropland. Malaj et al. (2020) identified a rapid systemic increase in chemical use and recommended adoption of policies to promote lower chemical reliance and sustainable agricultural practices.

5 Agriculture's image and change

Those who practice agriculture are or should be aware of its unfavorable public image. Agriculture had lost its connection to the public. Most people in developed countries are not farmers or ranchers and have no connection to the source of their food. We all must eat, but most people in the world's developed countries do not know any farmers or ranchers or how their food is produced.

However, for many modern, highly urbanized societies, farmers are a small sector of the economy. Only 1.3% (2.6 million Americans) are farmers and ranchers. In these societies, many, if not most, consumers may have no link to or knowledge of farming as an occupation and little understanding of how the food they eat is produced. Their experience of the countryside may be restricted to it being a place for recreation. This is the basis for potentially ill-informed attitudes of agricultural practices and farming policy (Wildavsky, 1995). This problem is well characterized by Kirschenmann (2000):

> If agriculture is purely an industrial act whose only purpose is to manipulate the technologies required to produce some wheat *with which we have no connection*, that is ground into flour in some distant factory *with which we have no connection*, made into frozen bread dough in some warehouse-like bakery *with which we have no connection*, and placed into a microwaveable plastic container *with which we have*

> *no connection* and all the while the process may be harming monarch butterflies, or rendering our water unfit to drink, or killing off the fish in our favorite streams *how could we expect the public to support agriculture?*

As in other areas of public policy, consumers, as citizens, have a responsibility to educate themselves so they can make informed decisions, whether it be the food choices they make, the government policies they support, or their vision of a sustainable future and how it can be achieved. Petrzelka et al. (1996) have argued for better education about public policy issues among citizens 'to raise their awareness of the need for increased research on the agronomic, economic, social, and moral issues of sustainable agriculture'.

When one attempts to determine who is responsible to change agriculture, it is clear the obligation is collective, not exclusively the responsibility of farmers and ranchers. It is not clear who will determine what has to be done and who will make sure it is done. Agriculture is a private enterprise with large public and environmental consequences. It stands to reason the public must be involved in creating a sustainable, environmentally correct agriculture with a firm moral foundation to ensure the continuation of human life on earth.

A more immediate picture of farms is provided by the United States Department of Agriculture (USDA) Farms and Land in Farms 2020 Summary compiled by the National Agricultural Statistics Service (NASS) (USDA-NASS 2021). There were 4400 fewer farms in 2020 compared to 2019. Total US land in farms was 896.6 million acres, a decrease of 800 000 acres from 2019. Fifty-one percent of US farms had less than $10 000 in sales in 2020. Producers reporting sales income of $10 000 to $99 999 operated 550 000 fewer acres than in 2019. Only 7.4% of US farms had sales of $500 000 or more but accounted for 40.8% of all farmland. According to the USDA Economic Research Service (ERS), in inflation-adjusted terms, net farm income in 2020 was slightly down compared to 2019 but 30.5% below its peak in 2013 (USDA-ERS 2021). In 2018, US farm income was 50% lower than 2013. These figures indicate the continuing drive toward larger, more capital input-intensive farming operations as well as the enduring economic pressures faced particularly by smaller-scale farmers in sustaining viable businesses (Thomas and Kevan, 1993). It is important to be aware that, while most US farmland is still owned by families and individuals, farmers may only be tenants or farm managers operating farms on behalf of landowners with their own targets for return on investment. Changes in farming practices may need to be negotiated and agreed by a range of parties. The market value of agricultural production is concentrated on fewer farms where more expensive technology has increased production and contributed to environmental and social problems. Increased capital requirements make it difficult to begin farming as farming and ranching became more costly and less profitable.

The ethics of agriculture and the ethical dimension of its many challenges have not been a primary or even a secondary concern within agriculture. All involved in agriculture know farming and ranching are important, virtuous endeavors worthy of continued public and governmental support. It is reasonable to claim people support agriculture even though they know little about it. The support may be for an idyllic vision of what agriculture used to be rather than what it is.

Because the grocery store is always full, there is little reason for the general public to be concerned about their food supply. With the exception of large farms, little government support has been provided to those who continue to farm. Many will argue the observed trends in agriculture follow the trends of consolidation and promotion of production efficiency in other important industries. If the country can be fed by some small number of large, highly efficient, well-managed farms and ranches, it will be good for all. It is a persuasive economic argument, not a moral argument.

The agricultural enterprise has not formulated and presented moral arguments to justify their practices because it was believed it had already won the moral case by feeding people, a morally acceptable achievement. Therefore, the moral and environmental challenges - the harm done - could be ignored because the benefits were greater than the costs. There are real, consequential public concerns about the negative effects of agriculture's technology and production practices. My conversations with farmers affirm they know the important problems: inappropriate pesticide use; harm to humans, the environment, and other species; loss of genetic diversity; disruption and loss of stable rural cultures; the fact that development policies favor large farms, small farmers being driven from the land; and development of crop monocultures with increased disease and insect susceptibility.

The negative effects, ignored by many agricultural scientists, agribusiness people, but not by all farmers, have been regarded as the price of bounty. Such unexamined, yet certain, environmental and moral positions are potentially dangerous to all engaged in agriculture. If the negative effects continue to be ignored and the hoped-for discussions and arguments about how food should be produced and agriculture's future sustainability do not occur, then those engaged in agriculture may lose without being engaged.

6 Public fear

Public fear about the harm caused by agricultural science has a moral foundation. People have a view of the future they expect, desire, or fear. Their view affects their risk perception which influences and determines government policy. But fear alone is not a reliable determinant of risk. A fear based on a perceived moral wrong (e.g. it is morally wrong to use pesticides that harm humans, non-target species, or the environment) and on the perception of harm is not the same as the presence of real harm. Irrational, scientifically unsupported fear is a powerful motivating force against agricultural practice. Wildavsky (1995) argued 'citizenship in science' as a prerequisite to moral outrage and any demands to stop an action or 'to get rid of the stuff', regardless of the cost. He advocates good science, but this is difficult because there is an appalling ignorance of science among the general public in countries such as the United States. What is wrong in Wildavsky's view is clear. Moral outrage has been allowed to lead policy in spite of clear scientific evidence not supporting the claim of harm to anything. Those who practice agriculture, those who do agricultural science, and those who raise moral issues and complaints must be morally and scientifically responsible and socially aware. Agricultural producers must continue to produce food and fiber and all concerned must recognize agriculture uses and cares for more land than any other human endeavor. Public concern about the real and unanticipated effects of the technology farmers use to produce food cannot be simply dismissed because it is not supported by scientific research. We have a right to know where our food comes from, ask questions about production, and expect answers. Too often, agriculture's response to public complaints and concern has been – Be quiet and listen, then you will understand why we do what we do.

Observations of and conversations with farmers confirm they want to farm so they can farm again tomorrow. Those who do research and develop agricultural technologies recognize the importance of economic survival. Conservation and sustainability practices may not be widely accepted because they do not allow farmers to demonstrate what they regard as symbols of good farming (Lavoie and Wardropper, 2021). As the Mali debate demonstrates, we can learn from the public's concern about what is done to some of our food.

Many people know little about science, are suspicious of unknown chemicals, and are concerned about their presence in their food. There is general agreement on the identity and use of best management practices (e.g. no-till farming, reduced pesticide use, and cover crops). On the other hand, there are significant gaps in knowledge between what some think farmers ought to do and why they farm as they do. Nearly all farmers, and all I have spoken to, agree they have a moral obligation to care for their land (Vaske et al., 2018). They consider themselves to be stewards of the land. They may not be able to quote Aldo Leopold's land ethic but they abide by his beliefs:

A thing is right when it tends to preserve the integrity, stability, and beauty of the biotic community. It is wrong when it tends otherwise.

Leopold saw the land as a community to which we belong, not as property where we can do what we want. We may use the land but only with love and respect. My view is most farmers agree.

There is widespread agreement among agricultural research scientists that cover crops[1] (alfalfa, rye, clovers, cowpeas, vetch, and several others) are beneficial to production. They reduce soil erosion, minimize nitrogen and pesticide leaching, and increase soil carbon storage. However, they are not widely used because cost-sharing is not available and research has not been translated into appropriate information for producers (Anderson-Wilk, 2008). Similarly, we know no-till agriculture greatly diminishes soil erosion, shields the ground surface, and retains soil moisture (Montgomery, 2008). Widespread adoption has not happened, although it could 'transform the relationship between humanity and the soil' (Montgomery, 2008). My conversations reveal farmers may not adopt what others regard as best management practices (e.g. crop rotation and no-till farming) because new expensive equipment may be required, their income is adequate, their problems are manageable, they are comfortable with the results of the practices they use, and they are not convinced new ways are better. In a very real sense, farmers recognize the value of the perfect but prefer to employ the possible.

7 Conservation

There has been little research by agricultural scientists on the role of values in determining what farmers and ranchers do. The independent variables consistently related to farmers' practices include self-identification as stewards of the land and non-financially motivated environmental attitudes. Conservation practices may not be widely accepted because they conflict with an experiential view of good farming, involve too much red tape for low payments, are expensive, and are seen as undesirable interference in farming operations. A survey of Iowa farm environmental leaders (Medina et al., 2020) was not representative of all farmers because it focused on leaders. It found most believed government conservation programs could be improved with more flexible program requirements and adjustments to payments. Some chose not to participate in government programs because they knew they were and would remain good stewards of the land.

In the European Union (EU), receipt of agricultural subsidies is increasingly predicated on delivery of public goods (Cusworth and Dodsworth, 2021). Societal expectations beyond food production are being imposed on agriculture. The changes are evidence that because those who practice agriculture have not been listened to, or participated in discussions of what is expected, social expectations have been imposed. The rules do not exempt agriculture from the obligation to produce good food. They impose obligations

[1] A cover crop is a non-cash crop planted to cover the soil, decrease erosion, improve soil fertility and quality, and aid control of weeds, insects, and diseases. They promote biodiversity of the agroecosystem.

including access to green spaces, healthy rural environments, flood resilience, and reduced greenhouse gas emissions.

My conversations demonstrate farmers value their land, community, good judgment, and the landscape. Nevertheless, conflicts occur between participants and program designers about the government program priorities concerning aesthetics, active land management, production, specific knowledge, community knowledge, agency over the landscape, and who pays (Chapman et al., 2019). Institutional structures of agri-environment programs frequently conflict with agrarian values and impede farmers' participation. Grounding conservation programs in 'locally salient values' might increase participation and foster stewardship consistent with societal requirements for conservation and ecological effectiveness. But only when increased costs are shared by those demanding the change and are not borne only by producers who are operating with a narrowing profit. Ranjan et al. (2019) found farmer attitudes toward conservation, environmental awareness, and trust in information sources strongly influenced adoption of conservation practices and programs. Individual farmers were influenced by economic and management needs and negative perceptions of conservation practices. Farmers often believed adoption of conservation practices was an economic risk. Improved practices developed by agricultural research and implemented by an agency have not been universally adopted by farmers in many nations, including the EU (Schmid et al., 2012), even with significant state financial support for adoption (Delaroche, 2020). Unfortunately, existing policies have considered landholders to be homogenous. It is tantamount to assuming a strawberry grower in eastern New York State, an Alabama cotton grower, an Illinois soybean grower, a Nebraska wheat grower, a Montana rancher, a Dutch farmer (See Runhaar et al., 2017), and an organic farmer in Iran (see Veisi et al., 2017) have identical values and motivation.

8 Sustainability

A problem with sustainability, something everyone is for, is there are no standards or norms to accurately assess the effects of the quest for sustainability on farming practices (Lampredi et al., 2019). It is generally agreed that agricultural practices must change to address climate change, biodiversity loss, and natural resource and land degradation. German agricultural policy has been affected by public demands for new regulations but has faced widespread protests from farmers (Gütschow et al., 2021). Farmers felt strongly restricted in their actions to implement diversified crop rotations to achieve sustainability. The problems most often cited were markets which severely limited the feasibility of growing some crops, limited regulatory predictability, and regulatory incoherence. The German experience emphasizes the need to create market and regulatory environments in which sustainable practices are accepted by and are profitable to farmers but not imposed on them.

The United Nations Department of Economic and Social Affairs developed 17 goals to achieve sustainable development. The 2030 agenda for sustainable development was adopted by all United Nation member states in 2015. It is a blueprint for peace and prosperity for people and the planet. The agenda recognizes ending poverty and other deprivations must be combined with improved health and education, reduced inequality, and economic growth – while tackling climate change and working to preserve oceans and forests. Moyer and Hedana (2020) developed a middle-of-the-road global development scenario. It includes 7 of the 17 goals, one of which is relevant to agriculture[2]. They suggest that without a significant shift in domestic and international aid policies and prioritization, it will be very difficult to achieve most of the development goals. At least 28 countries are especially vulnerable and are not projected to achieve any of the development goals. It is clear if agricultural sustainability is to be achieved it will have environmental effects. It is not equally clear that achieving it must also recognize the necessity of understanding what is required to improve and sustain the quality of life of farmers and ranchers who will bear the brunt of the responsibility to achieve sustainability (Bently Brymer et al., 2020). Farmers' well-being is 'experienced at individual and community level and influenced by ecological, economic, and socio-cultural processes'. It cannot be imposed by others.

Zeweld et al. (2017) claimed 'The introduction of sustainable practices is considered a win-win strategy for low-income countries' because it will improve food security and address environmental issues. They investigated views of small farmers on minimum tillage and row planting and found perceived usefulness. Efficacy, experience, social capital costs, and ease of operation were significant predictors of farmers' attitudes and acceptance. Information from the media did not affect farmers' intentions. They concluded adoption of sustainable practices would be improved when the advocates pay attention to social-psychological issues.

Over the past several decades, agricultural practices have resulted in significant environmental harm, loss of biodiversity, decreased water quality, and loss of ecosystem services. In Indiana, Grover and Gruver (2017) identified several barriers to adoption of sustainable farm management: markets, regulations, time and availability of labor, environmental and ecological factors, and access to education or research support. Farmers work in a complex system as they make decisions on how to farm. It is not sufficient to know and promote the advantages of systems developed by others to promote sustainable agriculture. Understanding farmers' attitudes and the reasons for their

[2] End hunger, achieve food security, improve nutrition, and promote sustainable agriculture.

participation or reluctance is vital. Grover and Gruver (2017) and Bowens (2015) strongly recommend 'local, small-scale agriculture should be approached as an integral part of economic development because it will be a vital component of a strong economy'. To achieve sustainability, the economic system must abandon the take-make-dispose principle and move toward a circular bio-based economy (Wensing et al., 2019). Studies of sustainable and alternative farmers and conventional farmers reveal the two groups have different beliefs, values, and attitudes about agriculture and rural life (Petrzelka et al., 1996). Achieving sustainability depends on a collective recognition of responsibility for the earth and future generations. Values will, as they always have, drive social change. Without study of and evaluation of farmers' motivations, it is doubtful if an environmentally sound global community will ever be achieved. For example, it is important, if not essential, to use easily obtained information about pesticide use and combine it with information supporting available practices unrelated to chemical use. Advocates of sustainability have had to depend primarily on their experience with sustainable practices because a lack of readily available research on sustainable agriculture has inhibited adoption. Petrzelka et al. (1996) claim public policy education is required if public and private institutions are to 'to raise their awareness of the need for increased research on the agronomic, economic, social, and moral issues of sustainable agriculture'. Research in New Zealand (Fairweather and Campbell, 2003) showed in contrast to conventional farmers, organic farmers had different attitudes about nature. They were moving toward sustainability, whereas farmers who adopted genetic engineering were committed to improving production. Agricultural research scientists and developers of agricultural technology assiduously avoid developing harmful farming technology. New technology does what it is designed to do. However, further research often shows it may cause undesirable effects (e.g. pesticides, gasoline engines, nitrogen fertilizer, and plowing). Research scientists and agricultural technologists always know what they are doing. Thinking about what they may be un-doing and what could go wrong is required.

The present capital, chemical, and energy-based system has not succeeded in feeding the world and has caused harm. The socioecological challenges and ethical dilemmas created by agricultural practice have led to increasing research on alternative farming systems as a desirable route to sustainability. If alternative agricultural systems had received the funding and emphasis the presently dominant system has received for several decades, we would be much closer to achieving agricultural sustainability.

An element of the quest for sustainability is the development of precision agriculture. It enthusiastically tells us how big data analytics, automated equipment, and decision-support software will optimize yields and ease inevitable public concern. A goal is to use artificial intelligence (AI) to improve the present system. Duncan et al. (2021) wonder if the goal of AI development is to create a smart agricultural economy to support the land and its stewards or if it is leading us down the same road and only providing support to development of AI systems.

The German experience, cited above, emphasizes the need to create stable market and regulatory environments in which sustainable practices can be profitably implemented by farmers but not imposed on them. In making the transition to more sustainable practices, many farmers perceive a trade-off between short-term production and profit and long-term environmental stability. This conflict between short- and long-term goals involves farmers, governments, and others promoting sustainable practices and how to allocate responsibility for implementing change and how it is funded (Chapman et al., 2019).

Roesch-McMally et al. (2017) emphasized the importance of strengthening the existing, enduring ethic of stewardship among farmers to prioritize the longer-term stability of farming operations as farmers navigate the transition between short-term production and long-term conservation and sustainability goals. This will require improved flexibility and payment schemes (Medina et al., 2020).

My conversations with farmers strongly suggest constructive engagement between farmers and conservationists will contribute to long-term commitment by farmers to practices to achieve sustainability and affect farmers' attitudes to environmental issues (Riley, 2016). The potential of a participatory approach to promote sustainable agricultural practices has been demonstrated when the effects of agri-environmental proposals include the respective interests of policymakers or regulators, local communities, and farmers (Mauchline et al., 2012). A participatory approach makes it possible to establish and rank agreed targets in a transparent and credible way that is accepted by all stakeholders including farmers. Early involvement of farmers and other stakeholders will provide a way to give farmers both the understanding and mechanisms to engage and build consensus within society necessary to build a sustainable future for all (Cullen et al., 2018).

9 References

Anderson-Wilk, M. 2008. The gap between cover crop knowledge and practice. *Journal of Soil and Water Conservation* 63(4):96A.

Andrée, P. 2021. Hugh Campbell: Farming inside invisible worlds: Modernist agriculture and its consequences. *Agriculture and Human Values* 38(4):1223–1224.

Benefeldt, E. S., McGonagle, M. and Niewolny, K. 2021. Rethinking farmer knowledge from farm soil to plate through narrative inquiry: And Aguilar food systems perspective. *Journal of Agriculture, Food Systems, and Community Development* 11(1):1–15. doi: 10.5304/jafscd.2021.111.012.

Bently Brymer, A. L., Toledo, D., Spiegal, S., Pierson, F., Clark, P. E. and Wulfhorst, J. D. 2020. Social-ecological processes and impacts affect individual and social well-being in a rural Western U.S. landscape. *Frontiers in Sustainable Food Systems* 4:Article 38.

Bowens, N. 2015. *The Color of Food - Stories of Race, Resilience and Farming*. New Society Publishers got Biola Island, Canada, 228 pp.

Campbell, H. and J. R. Fairweather 1998. The Development of Organic Horticultural Exports in the Development of Organic Horticultural Exports in New Zealand. Research report number 238. Agribusiness and Economics Research Unit. Lincoln University, Canterbury, New Zealand, 70 pp.

Chapman, M., Satterfield, T. and Chan, K. M. A. 2019. When value conflicts are barriers: Can relational values help explain farmer participation in conservation incentive programs? *Land Use Policy* 82:464–475.

Coon, J. J., Van Riper, C. J., Morton, L. W. and Miller, J. R. 2020. What drives private landowner decisions? Exploring non-native grass management in the Eastern Great Plains. *Journal of Environmental Management* 276:111355.

Cullen, P., Dupraz, P., Moran, J., Murphy, P., O'Flaherty, O'Donoghue, C., O'Shea, R. and Ryan, M. 2018. Agri-environment Scheme design: Past lessons and future design. *EuroChoices* 17(7):1–5.

Cusworth, G. and Dodsworth, J. 2021. Using the 'good farmer' concept to explore agricultural attitudes to the provision of public goods. A case study of participants in an English agri-environment scheme. *Agriculture and Human Values* 38(4):929–941.

Delaroche, M. 2020. Adoption of conservation practices: What have we learned from two decades of social-psychological approaches? *Current Opinion in Environmental Sustainability* 45:25–35.

Duncan, E., Glaros, A., Ross, D. Z. and Nost, E. 2021. New but for whom? Discourses of innovation in precision agriculture. *Agriculture and Human Values* 38(4):1181–1199.

Ehsan, E., Zhang, H., Lirong, X., Khalid, Z. and Xu, H. 2021. Understanding cognitive and socio-psychological factors determining farmers intentions to use improved grassland: Implications of land-use policy for sustainable pasture production. *Land Use Policy* 102:105250.

Emery, S. B. and Franks, J. R. 2012. The potential for collaborative agri-environmental schemes in England: Can a well-designed collaborative approach address farmers' concerns with existing schemes? *Journal of Rural Studies* 28(3):218–231.

Fairweather, J. R. and Campbell, H. R. 2003. Environmental beliefs and farm practices of New Zealand farmers: Contrasting pathways to sustainability. *Agriculture and Human Values* 20(3):287–300.

Gremmen, B. 2020. An ethical compass for agriculture. Proceedings of the 3rd International Congress on Agricultural and Food Ethics. Istanbul, Turkey, pp. 31–36.

Grover, S. and Gruver, J. 2017. 'Slow to change': Farmers perceptions of place-based barriers to this chapter now from the farmers sustainable agriculture. *Renewable Agriculture and Food Systems* 32(6):511–523.

Gütschow, M., Bartkowski, B. and Filipe-Lucia, M. R. 2021. Farmers' action space to adopt sustainable practices: A study of arable farming in Saxony. *Regional Environmental Change* 21(4):103.

Kirschenmann, F. 2000. *Challenges Facing Philosophy as We Enter the 21st Century: Reshaping the Way the Human Species Feeds Itself*. The Eddy Lecture, Colorado State University, 24pp. Available from Colorado State University Department of Philosophy.

Lampredi, M. G., Sorensen, C. G. and Bochtis, D. 2019. Agricultural sustainability: A review of concepts and methods. *Sustainability* 11(18):5120, 27 pp.

Lavoie, A. and Wardropper, C. B. 2021. Engagement with conservation tillage shaped by "good farmer" identity. *Agriculture and Human Values* 38(4):975-985.

Leopold, A. 1966. A Sand County Almanac. Original publication 1949, Oxford University Press.

Malaj, E., Freistadt, L. and Morrissey, C. A. 2020. Spatio-temporal patterns of crops and agrochemicals in Canada over 35 years. *Frontiers in Environmental Science* 8:Article 566452.

Mauchline, A. L., Mortimer, S. R., Park, J. R., Finn, J. A., Haysom, K., Westbury, D. B., Purvis, G., Louwagie, G., Northey, G., Primdahl, J., Vejre, H., Kristensen, L. S., Teilmann, K. V., Vesterager, J. P., Knickel, K., Kasperczyk, N., Balázs, K., Podmaniczky, L., Vlahos, G., Christopoulos, S., Kröger, L., Aakkula, J. and Yli-Viikari, A. 2012. Environmental evaluation of agri-environmental schemes using participatory approaches: Experiences of testing the agri-environment Footprint Index. *Land Use Policy* 29(2):317-328.

Mayberry, D., Crase, L. and Gullifer, C. 2005. Categorizing farming values as economic, conservation and lifestyle. *Journal of Economic Psychology* 26(1):59-72.

Medina, G., Isley, C. and Arbuckle, J. 2020. Iowa farm environmental leaders' perspectives on the U. S. Farm Bill conservation programs. *Environment, Development and Sustainability* 23:173-194.

Montgomery, D. R. 2008. Agriculture's no-till revolution? *Journal of Soil and Water Conservation* 63(3):64A-65A.

Moyer, J. D. and Hedana, S. 2020. Are we on the right path to achieve the sustainable development goals? *World Development* 127:104749.

National Research Council. 1989. Problems in U. S. agriculture. In: *Alternative Agriculture*, pp. 89-134 (esp. pp. 90-93). National Academy of Science Press, Washington, DC.

Nleya, T. and Clay, S. 2021. Near-term problems in meeting world food demands at regional levels: A special overview. *Agronomy Journal* 6 pp. doi:10.1002/agj2.20903.

Perry, V. and Davenport, M. A. 2020. An inductive framework of self-efficacy to understand and support farmers and conservation agriculture. *Journal of Soil and Water Conservation* 75(2):198-208.

Petrzelka, P., Korsching, P. F. and Malia, J. E. 1996. Farmers attitudes and behavior toward sustainable agriculture. *The Journal of Environmental Education* 28(1):38-44.

Pimbert, M. P. and Barry, B. 2021. Let the people decide: Citizen deliberation on the role of GMO's in Mali's agriculture. *Agriculture and Human Values* 38(4):1097-1122.

Prokopy, L. S., Flores, K., Arbuckle, J. G., Church, S. P., Eanes, F. R., Gao, Y., Gramig, B. M., Ranjan, P. and Singh, A. S. 2019. Adoption of agricultural conservation practices in the United States: Evidence from 35 years of quantitative literature. *Journal of Soil and Water Conservation* 74(5):520-534.

Prokopy, L. S., Floress, K., Klotthor-Weinkauf, D. and Baumgart-Getz, A. 2008. Determinants of agricultural best management practice adoption: Evidence from the literature. *Journal of Soil and Water Conservation* 63(5):300-311.

Ranjan, P., Church, S. P., Floress, K. and Prokopy, L. S. 2019. Synthesizing conservation motivations and barriers: What have we learned from qualitative studies of farmers behaviors in the United States? *Society and Natural Resources* 32(11):1171-1199.

Riley, M. 2016. How does longer-term participation in agri-environment schemes (re) shape farmers' environmental dispositions and identities? *Land Use Policy* 52:62075.

Roesch-McMally, G., Arbuckle, G. and Tyndall, J. C. 2017. Soil as social-ecological feedback: Examining the ethic of soil stewardship among corn belt farmers. *Rural Sociology* 83: 145–173.

Rossi, A. M. and Hinrichs, C. C. 2011. Hope and skepticism: Farmer and local community views on the socio-economic benefits of agricultural bioenergy. *Biomass and Bioenergy* 35(4):1418–1428.

Runhaar, H. A. C., Melman, T. C. P., Boonstra, F. G., Erisman, J. W., Horlings, L. G., de Snoo, G. R., Termeer, C. J. A. M., Wassen, M. J., Westerink, J. and Arts, B. J. M. 2017. Promoting nature conversation by Dutch farmers: A governance perspective. *International Journal of Agricultural Sustainability* 15(3):264–281.

Schmid, O., Padel, S. and Levidow, L. 2012. The bio-economy concept and knowledge base in a public goods and farmer perspective. *Bio-Based and Applied Economics* 1(1):47–63.

Spangler, K., Biuchfield, E. K. and Schumacher, B. 2020. Past and current dynamics of U.S. Agricultural land use and policy. *Frontiers in Sustainable Food Systems* 4:Article 98.

Thomas, V. G. and Kevan, P. G. 1993. Basic principles of agroecology and sustainable agriculture. *Journal of Agricultural and Environmental Ethics* 6:1–9.

Thompson, P. B. 2020. Agriculture and food ethics: Is it time to go back to basics? Proceedings of the - of the 3rd Turkish Congress on Agriculture and Food Ethics, Istanbul, Turkey.

United Nations, Human Rights Council. 2017. Report of the special rapporteur on the on the right to food. Presented to the UN Commission on Human Rights, A/HRC/34/48, 24pp.

USDA-ERS. 2021a. *United States Department of Agriculture – Economic Research Service (USDA-ERS): Farming and Farm Income*. Available at: https://www.ers.usda.gov/data-products/ag-and-food-statistics-charting-the-essentials/farming-and-farm-income/.

USDA-NASS. 2021b. *United States Department of Agriculture – National Agricultural Statistics Service (USDA-NASS): Farms and Land in Farms 2020 Summary*, pp. 76–82. Available at: https://www.nass.usda.gov/Publications/Todays_Reports/reports/fnlo0221.pdf (Accessed April 2022).

Vaske, J. J., Miller, C. A., Toombs, T. P., Schweizer, L. A. and Powlen, K. A. 2018. Farmers' value orientations, property rights and responsibilities, and willingness to adopt Leopold's land ethic. *Society and Natural Resources* 31(10):1118–1131.

Veisi, H., Carolan, M. S. and Alipour, A. 2017. Exploring the motivations and problems of farmers for conversion to organic farming in Iran. *International Journal of Agricultural Sustainability* 15(3):303–320.

Wensing, J., Carraresi, L. and Bröring, S. 2019. Do pro-environmental values, beliefs and norms drive farmers interest in novel practices fostering the bio economy? *Journal of Environmental Management* 232:858–867.

Wildavsky, A. 1995. *But Is It True? A Citizens Guide to Environmental and Health and Safety Issues*. Harvard University Press, Cambridge, MA, 547 pp.

Zeweld, W., Van Huylenbroeck, G., Tesfay, G. and Speelman, S. 2017. Smallholder farmers' behavioural intentions toward sustainable agricultural practices. *Journal of Environmental Management* 187:71–81.

Chapter 6

Virtue ethics, Wendell Berry and agriculture

Wade Casey, Loyola University Chicago, USA

1 Introduction

What do virtues have to do with agriculture? What are the traits possessed by virtuous agricultural workers or a virtuous agricultural sector? How might such agriculture alleviate strains on people, animals, and land? How might the embodiment of virtues sustain the world and allow it to flourish? And what sort of virtues would make this possible? This chapter seeks answers to these questions or at least attempts to move conversations surrounding agricultural ethics toward answers to them. It is the hope of this chapter that through an exploration of virtue ethics, the agricultural community will be better positioned to allow all things on our planet to thrive.

The chapter will begin with a brief overview of the virtue ethics tradition. I will detail the basic contours of this heritage, and how it is that it remains a vital resource in our contemporary climate. I will then attend to Aristotle's notion of virtue which serves as perhaps the most foundational resource for virtue ethics in the West. With this survey of Western virtue ethics in mind, I will turn to the writings of Wendell Berry to suggest the sort of agrarian principles Berry writes about and embodies are excellent, and importantly effective, renditions of virtue ethics for modern agriculture. I will briefly introduce Berry to any audience less familiar with his writings, and then I will consider the ways in which the agricultural virtues he advocates for serve to advance planetary well-being on both a local and global scale. Finally, I will conclude by offering a few recommendations for agriculture in light of my reflections on virtue ethics and Berry.

http://dx.doi.org/10.19103/AS.2023.0125.16

2 Introducing virtue ethics

Virtue ethics enjoys a time-honored place within the history of ethical traditions. In the West, Plato served as a sort of torchbearer of virtue ethics. However, it was his student Aristotle, to quote Alasdair MacIntyre (1981, p. 146), who was the tradition's 'greatest representative'. The development of Greek virtue ethics certainly dates back much further than Plato and Aristotle, for one can observe the basic framework for Plato's ideas about a virtuous society in Homer's *Iliad* and *Odyssey*. For Homer, as picked up later by Plato, the world is ordered in a particular way, with each person on Earth serving a specific social role. Some virtues in Plato's mind are utility virtues, meaning Plato (*The Republic*, p. 4) understands them to pertain to all levels of society - i.e. justice and moderation - whereas other virtues Plato believed should only (or mostly) apply to specific groups - i.e. courage for soldiers. Without doubt, the ideal of virtuous living was not a product only of the Western philosophical world. Eastern figures such as Confucius reinvigorated interest and commitment to virtuous ideals during the latter part of the Zhou dynasty (eleventh to eighth centuries BCE), a violent period in which compassion was lacking at many levels of society.[1] However, given my background in Western philosophy, much of what I say in this chapter builds on the Western intellectual heritage.

Virtue ethics is one of the dominant contemporary moral theories among philosophers and ethicists. The other two preeminent theories, over the past couple hundred years in the Western world, are consequentialism and deontology - the former often associated with Jeremy Bentham and John Stuart Mill and the latter with Immanuel Kant. As Matt Zwolinski and David Schmidtz (2013, p. 221) argue, virtue ethics differs from these other two traditions in that virtue ethics asks a very different question. These former two theories are chiefly interested in what makes an action right, whereas virtue ethics concerns itself with what sort of life a person is to live.

> Often associated with Aristotle, but with roots in various traditions . . . virtue ethics tells us that what is right is to be a certain kind of person, a person of virtue: courageous, modest, honest, even-handed, industrious, wise. A virtuous person will, of course, express his or her virtue through action. But, for virtue ethics, the specification of rules of right action is largely a secondary matter - one that in many ways presupposes the kind of practical wisdom possessed by the person of virtue. (Zwolinski and Schimdtz, 2013)

In other words, virtue ethics concerns itself with the formation of the human person and the kinds of characteristics that allow people and the world as a whole to flourish. This means that virtue ethics cares about what sort of person each of us is and what sort of *character* enables us to be good. As Aristotle

[1] For an excellent consideration of Confucian virtue ethics, see the works of Philip J. Ivanhoe.

(*Nicomachean Ethics*, 2.4, 1105a30-35) says, a virtuous person does not care only for the result of an action but how that action was achieved.

As Daniel C. Russell makes clear, virtue ethics' emphasis on character formation aims at enabling a person and groups of people to live lives that are meaningful and happy. Only when the goal of happiness is squarely in one's focus do questions of right and wrong find their appropriate place (Russell, 2013b, p. 7). In the virtue ethics tradition, as I will show below in my evaluation of Aristotle, happiness is not the ephemeral feeling modern people often associate with the term. Rather, happiness is about well-being, a life lived in a fulfilling manner. For Berry and agricultural ethics, happiness must include not only human fulfillment and well-being but also the fulfillment of the whole material world, a pursuit that requires humanity to find its harmonious place within the whole. As Russell (2013b, p. 7) says, 'practical reasoning [or reasoning dealing with what is the best way to live] leads to thought about the virtues, excellences of character that consist in both caring about the right sorts of things and having the wisdom and practical skills to judge and act successfully with respect to those things'. Though the basics of agricultural virtue ethics are already taking form, let me first turn to explicate the ideas of the good, happiness, and virtue in the thought of the 'greatest representative' of the tradition.

2.1 Aristotle

If one wants to understand the development of the virtue ethics tradition, there is no more appropriate place to begin than with Aristotle and his *Nicomachean Ethics*.[2] Plato, Socrates, and the pre-Socratics made it evident that particular virtues were to be understood as necessary components of ethical living for those residing in the *polis*. Aristotle affirmed this but wanted to emphasize that virtues are in the service of such ethical living because they form in a person the kind of character that makes that person the best possible version of oneself. An ethical system built on virtues - at least as Aristotle conceived of it - is teleological in nature, meaning that practicing virtues carry with them a distinct aim or purpose. As Aristotle (*NE*, 1.1, 1094a1-3) says to open the work, '[e]very art and applied science and every systemic investigation, and similarly every action and choice, seem to aim at some good; the good, therefore, has been well defined as that at which all things aim'. As he (*NE*, 1.1, 1094a6-9) explains with examples: a doctor's aim is health, a shipbuilder aims at constructing a well-built vessel, and a household aims at financial stability

[2] Referenced as *NE* throughout the remainder of the chapter. For a good introduction to Aristotle's other, though slightly less extensive, work on ethics see Brad Inwood and Raphael Woolf (Eds) (2013), *Aristotle: Eudemian Ethics*, Cambridge University Press, Cambridge, UK; Guilio Di Basilio (Ed) (2022), *Investigating the Relationship between Aristotle's Eudemian Ethics and Nicomachean Ethics*, Routledge, London, UK.

and prosperity. These are particular goods for particular types of people, with particular professions or in particular stages of life.

But thinking beyond particularities, what might be said of the aim or goal of virtue in a general sense? Aristotle (*NE*, 1.4, 1095a19) believes virtues form in a person a disposition, or a sort of character, that enables said person to move toward the general human good which he calls happiness, or eudaimonia. As noted above, according to Aristotle (*NE*, 1.10, 1100b19), happiness is not some fleeting or whimsical good feeling but rather a sort of stability, or an 'attribute of permanence'. Again, to quote MacIntyre (1981, p. 148), 'it is a state of being well and doing well in being well'. Happiness is the ultimate good of human life because it is that thing to which all other things lead. One might rightly suggest that financial stability, health, honor, close and trustworthy friends and family (and so on) are also goods. But Aristotle believes that these goods' ultimate purpose and trajectory is happiness, whereas happiness does not inevitably lead to these goods. Thus, eudaimonia is the highest human good and the only thing that he considers to be a good for its own sake, and thus, as an end in itself.

After laying out his ideas about human purpose and happiness at the beginning of the work, Aristotle devotes the majority of Book II of the *NE* to articulating his understanding of virtue. Simply stated, virtue is a characteristic that involves a choice and that, when employed well, helps a person find the middle point between two extremes, or what the philosopher (*NE*, 2.6, 1107a1) calls the 'mean'. The mean lies between the extremes of excess and deficiency/ defect. For example (*NE*, 4.1, 1119b20-35), generosity can be considered a virtue because it lies between the excessive extreme of extravagance and its deficient opposite, stinginess. Or take one of the cardinal virtues Plato popularized, courage. Aristotle (*NE*, 3.7, 1115b7-1116a9) says that courage is the virtuous mean between the excess of recklessness (or rashness) and the deficiency of cowardice.

Though these middle points in these instances might be obvious, Aristotle does not believe finding the mean between the vices of excess and deficiency is always an easy task. This is where intellectual virtues make their entrance. Aristotle (*NE*, 1.13, 1103a4-10; *NE*, 6) differentiates between intellectual virtues (i.e. wisdom, understanding, expertise) and moral virtues (i.e. generosity, self-restraint).[3] What makes the mind and reason so important for virtuousness is that 'it is the intellectual virtue without which none of the other virtues of character can be practiced' (MacIntyre, 1981, p. 154). Think back to the quotation of Dan Russell in the introduction: for one to be virtuous, one must possess 'the

[3] This is to say nothing of Aristotle's distinction between practical or political happiness and theoretical happiness. For a good overview of this distinction, see Pierre Hadot (2002), *What Is Ancient Philosophy?*, Belknap Press, Cambridge, MA, 78ff; for one the seminal considerations of Aristotle's ethics see Sarah Broadie (1991), *Ethics with Aristotle*, Oxford University Press, Oxford, UK. Broadie specifically addresses the significance of 'practice' and 'theory' in Chapters 4 and 7.

wisdom and practical skills to judge and act successfully with respect to [the virtues]'. Moral virtues enable us to live rightly. But to make the right choices, particularly when those choices are difficult, one needs the requisite knowledge *and* wisdom to act morally.

The importance of highlighting this final point cannot be overstated. Aristotle is convinced that for a person to practice virtues like moderation, gentleness, or justice, one must possess the knowledge of a virtue and the wisdom to employ it properly. With this account of Aristotle's notion of virtue laid out, it is now time to put it in conversation with Wendell Berry's agricultural ethics. I will not only show the vitality of moral virtues for Berry's agricultural ethics but, to this final point just discussed, show how the acquisition of knowledge plays a central, though albeit unconventional, role in his ethical framework.

3 Wendell Berry and a virtuous agriculture

For more than 60 years, Wendell Berry has written extensively about ecological and social matters. His writings have covered topics ranging from war to abortion to race to technology. But no single topic occupies Berry's attention more than agriculture and particularly the ways he believes industrialized agriculture harms both natural environments and the human communities who depend on them. Berry often wields an impressive knowledge of issues related to agriculture and ecology in his writings. But his acumen in such matters is not simply an armchair perspective but rather one that is thoroughly informed and conditioned by what he (Berry, 1981a) calls his day job: 'subsistence farming'. Berry (2017, p. 223) uses this term to describe a way of life in which one lives on, with, and from the land.

Berry (b. 1934) grew up in rural Henry County, Kentucky. He is the eldest of four children born to a multi-generational farming family. His father farmed tobacco and later became an attorney to help advocate for their small tobacco program, Burley Tobacco Growers Cooperative Association (Berry, 2012b, p. 44). Early in Berry's adulthood, he moved away from farming and toward a life in the academy, earning a BA and MA from the University of Kentucky in English. In 1958 he attended Stanford University's creative writing program as a Wallace Stegner Fellow. A few years later, he received a Guggenheim Fellowship, followed by teaching stints at New York University (1962–1964) and the University of Kentucky (1964–1977). He published his first novel *Nathan Coulter* in 1960 and his first non-fiction work *The Long-Legged House* in 1965.

Berry could be appropriately utilized by any number of writers interested in agricultural ethics. However, in this present chapter, I choose to read him alongside the virtue ethics tradition for precisely this reason: core to Berry's solutions to a host of agricultural issues is his consuming interest in the character of the human person. Berry's second essay in his widely popular *The Unsettling of America: Culture and Agriculture* is titled 'The Ecological Crisis as a Crisis of Character'. In fact, one would be hard-pressed to pick up any work by Berry and not find at least some passing comment about the significance of how people live their lives and how such lives lived impact the world around them. Let me turn now to explore the virtuous shape of his thought.

3.1 Humility

From Berry's earliest writings, he reveals his conviction that ecology roots itself in the notion that all things on Earth depend on each other and form 'an interlocking system' of support. Though humans understand aspects of this system, Berry (2002, p. 89) emphasized in one of his seminal early essays 'Think Little' that humans will never fully understand this system nor will they ever fully control it.[4] More than 40 years later, Berry (2015, p. 46) reemphasized this point in an essay about forestry and logging: 'The connections and interactions among all the creatures in a thriving forest ecosystem are complex beyond the human ability to think. This is the starting point, the primary axiom [for ecology]'. Certainly, Berry's early rendition of this thought reveals that all ecosystems (not just forests) 'are complex beyond the human ability to think'. Thus, when one irrigates corn on the plains or raises cattle in the mountains, it is the responsible and necessary thing to think and act without hubris, in ways that take note of and show care for such complexities.

How might one begin down this anti-arrogance, agricultural road? Berry starts by highlighting the importance of humility as a foundational virtue for agricultural responsibility in light of the immense complexities of the natural world. Berry (2015, p. 46) says: 'humility is the primary virtue of good forestry. One must get the scale right, so as not to put too much at risk. One must not use too much power, or be in too much of a hurry'. This is the very point Berry makes in 'Think Little', when he says, 'our great dangerousness is that, locked in our selfish and myopic economics, we have been willing to change or destroy far beyond our power to understand'. Why is it that Berry believes humans and their economic systems have become this way? Because we have not been humble or reverent enough toward what we do not know and what we cannot control (Berry, 2002, p. 89).

These ideas fit squarely into Berry's notion of limits. For many forms of agriculture to be practiced well, Berry believes one must appreciate concerns like the carrying capacity of the land (Berry, 2017, p. 55; Berry, 1981b, pp. 134–145) and localized knowledge which understands best practices for particular times and places (Berry, 2017, p. 38; Berry, 1965). When one intimately knows

[4] 'Think Little' (1970) appeared in several publications prior to its inclusion in The Art of the Commonplace.

the land, animals, and plants, one recognizes the value of treating all with care and respect to accomplish the best possible outcome. No responsible farmer, rancher, or forester treats the source(s) of their existence in a manner that might destroy it, given their economic well-being depends on proper care and productivity of the source.

These ideas harken back to Aristotle's notion of the mean. Think of Aristotle's example of the good shipbuilder. For a shipbuilder to find the right balance between crafting a ship too quickly and haphazardly or not quickly enough, the builder must take proper precautions but also be motivated by economics. To strike this balance, the builder must work with the right lumber, tools, and measurements. Without proper care in building, the ship might break apart in a strong storm or sink to the bottom of the sea. However, if the builder does not work quickly enough, then ships might never be built and the builder could not make a living because sailors could not sail.

This embodies many of the valuable aspects of Berry's emphasis on good farming that recognizes the limits of a resource but also the necessity of being able to live on account of the resource. Finding the middle point between excessive or exhaustive agriculture and deficient agriculture is what separates good farmers and ranchers from bad ones. In Berry's mind, a virtue like humility encourages a farmer to acquire extensive knowledge about places, animals, seasons, equipment, etc. But this humility also prevents a farmer from ever believing one has absolute knowledge of one's place and one's craft.

Throughout Berry's writings, he laments that localized agricultural knowledge and skill have been consistently replaced by an industrial agricultural complex with its value of large-scale, one-size-fits-all, remotely controlled practices. Too frequently, big corporations are driven by maximizing production and profits and thus care little about making choices that are wise for particular ecosystems and the human communities who depended on them. In such an agricultural environment, there is no time to think through best practices for the longevity of local lands, animals, and people because high outputs are needed now – even if such returns result in a jeopardized, perilous future.

3.2 Generosity

Berry's concerns over modern agriculture's insatiable drive for higher yields and greater profits necessitate a second agricultural virtue he calls generosity. By generosity, Berry means that a person engaged in healthy agricultural practices does not treat nature selfishly or greedily. Rather, 'a good forester thinks first of the forest', as Berry (2015, p. 47) says. Berry quotes a forester named Troy, from whom Berry gleaned principles of good forestry. Troy (Berry, 2015, p. 47) told Berry, 'A bad logger goes to the woods thinking what he can take out. A good logger goes to the woods thinking of what he should leave'. Such an idea flies in the face of so much of the US economy, not just agriculture. And yet, agricultural enterprises seem to have a unique power to harm natural environments like forests or aquifers that when treated poorly, take hundreds or even thousands of years to rejuvenate, if they do so at all.

Berry rightly perceives much of contemporary agriculture to be an enterprise beholden to the maximum efficiency, maximum gain framework. Whatever practice produces the highest yield with the least amount of input is the preferable procedure. Berry knows full well the human necessity of agricultural production for an economic environment that enables human survival. But Berry also understands that ecological and agricultural practices that serve economic ends need not seek those ends in ways that deplete or ruin the source.

When Berry speaks of generosity toward natural resources, as a means to counteract selfish and greedy tendencies, he also highlights the value of the moral virtue of temperance or self-restraint. According to Berry, self-restraint functions not simply as an austere buzz-kill or a sort of environmental self-flagellation. Rather, its purpose is to honor the limits of a resource and to protect its integrity not only for potential future use but also as an inherent good in its current state. In this way, Berry considers the natural world to be a place of infinite wonder and beauty, a place that sustains human life when cared for well but one that can be pushed too far.

The enormous amounts of waste produced through agricultural industrialism indicate to Berry that too often selfishness and greed govern agriculture as well as much of the global economy. Berry (2002, p. 83) calls waste the 'first principle' of the US economy, and he rightly notes that when economic prosperity is more limited (or not present at all), principles of thrift and reuse mark the economics of most humans on earth. Berry (2017b, p. 23) suggests such principles were much more common in American agricultural life prior to WWII, after which he believes industrialization across America's farmlands burgeoned, leading to extraordinary increases in production levels that carried with them commensurate increases in waste and pollution.

Undoubtedly, industrial agriculture launched the United States into an unprecedented position of global agricultural dominance. This dominance was fully appreciated by many large-scale farmers and ranchers as well as many adjacent executives working for chemical, energy, and equipment companies (to name a few). As such, wastefulness and ecological degradation have been overshadowed by the glimmer of riches few humans in history have known. And yet, these riches have impoverished our planet and many who live on it. It appears it is only a matter of time before such an impoverished world will not be able to maintain our wastefulness and destruction, a conclusion that says little of the planet's current desire to do so.

Berry's concern over waste and degradation, I believe, is core to his emphasis on both humility and generosity. Unfortunately, humans possess a propensity to view their well-being (or perceived path toward well-being) as being more valuable than the well-being of other people as well as all other material things on the planet. This mentality views well-being as zero-sum. And though a sort of Darwinian survival of the fittest conditions all of life on earth, there is also an equally powerful force of mutuality shared by all things on earth, one that honors the natural regenerative cycle of life, growth, death, decay, life, etc.

Berry's emphasis on virtues, character formation, and economics of reuse and thrift offers a corrective to such a zero-sum framework. It does not deny the drive to survive but suggests that we as humans desire more than simple survival. We also desire to thrive, and our abilities to achieve a life filled with flourishing necessitate mutuality that cares for and respects the life and well-being of things beyond our own experiences and knowledge. Such an effort, Berry recognizes, is difficult. To learn to operate in such a manner, in agriculture and beyond, Berry thinks humans must develop sympathies that encourage a more holistic, universal care for our worlds (both big and small). It is to the development of such 'sympathies' that I now turn to as I evaluate one final virtue in Berry's thought.

3.3 Love

This final virtue from Berry's agricultural writings is one he (2015, p. 51) acknowledges as a difficult term in agricultural conversations, one that might even cause some to be 'squeamish'. The virtue is love or affection. The role of this virtue cannot be underestimated in Berry's account of the way the world should be, and it is seemingly the orienting feature of so much of his agricultural framework. As Jeffrey Bilbro (2019, p. 7) says, 'if the industrial economy turns on explosive violence, Berry's sustainable economy turns on affection'.[5] Bilbro (2019, p. 17) explains that the virtues of such an economy are renewal oriented, meaning they honor the reuse emphases mentioned earlier. Rather than 'working against' nature, such virtues of renewal strive to 'work with' the natural world. In a beautiful passage that embodies well Berry's spirit of renewal, Bilbro (2019, p. 20) writes: 'Rather than developing techniques to make our places more easily meet our individual desires, members of sustainable economies learn to die to their own desires and develop the virtues needed to serve the needs of their beloved places'.

In addition to Berry's concern that discussions of love in agriculture can cause squeamishness, he notes that many think such serious conversations (about

[5] Bilbro echoes the title of Berry's delivered speech at 41st Annual Jefferson Lecture of the National Endowment for the Humanities 'It All Turns on Affection'.

economics, agriculture, etc.) should avoid consideration of one's affections. As Berry (2012a, pp. 14–15) says, 'The charge will be made that affection is an emotion, merely "subjective", and therefore that all affections are more or less equal: people may have affections for their children and their automobiles, their neighbors and their weapons'. But the subjective, emotional nature of affection seems not to be the reason Berry believes many agriculturalists, economists, academics, business executives, and politicians reject consideration of the virtue. Rather, he thinks what causes such consternation for people when discussions of affection arise is that a virtue like love demands commitment on a personal level. And this matters tremendously, as Berry (2012a, p. 15) says because it is nothing if it is not personal.

What I take Berry to mean here is that if people concerned about any topic do not take a personal interest and stake in the subject matter, there is a much greater chance the outcome of that conversation, the publications written about it and the policy and laws that result from it, will leave too much to abstraction. Abstraction will enable the continued distance between those discussing/writing about/legislating for the topic (in our case, agriculture) and those performing the work *and* the land and animals affected by that work. Berry believes that for agricultural work to be done well, one needs to remain close enough to the work so that one might respond appropriately.

Berry's call to personal responsibility is not his attempt to diminish the significance of public and institutional responsibility. But Berry believes that no public or institutional response to a crisis will be effective without personal and private convictions and actions: 'that there is no public crisis that is not also private' (Berry, 2002, p. 82). Core to Berry's conviction that affection begins in private is his notion of imagining oneself close to and connected to the thing, place, or person one loves. 'For humans to have a responsible relationship to the world, they must imagine their place in it. To have a place, to live and belong in a place, to live from a place without destroying it, we must imagine it' (Berry, 2012a, p. 14). Joseph Wiebe (2017) has written excellently about this emphasis in Berry's thought. My interest in bringing attention to it here is simply to highlight that Berry does not believe one can love and thus adequately care for people, places, or things that one is not close to because one cannot have intimate knowledge of said things without this closeness.

The emphasis on intimate knowledge brings us back to the earlier discussion of humility. Berry believes that humility is the requisite virtue for creatures like ourselves who have limited knowledge of the complexities of our world. And though we can know important things about our world, we really can only know such things (about the intricacies of land, plants, animals, and water) through close observation and attention to them. When we properly recognize these complexities, we are humbled and Berry believes generosity takes root in our thoughts and actions. This spirit of generosity enables us to not view things in a purely self-interested or greedy manner. Rather, we are better prepared to consider the whole, the whole which we do not fully understand, and act in ways that respect our beautifully complex world. But to be positioned to act with such respect and care, we must be motivated by love. Only when our affections lead the way, only when we are motivated by a desire to improve the whole world rather than simply our own lives do we, as a species, stand a chance to play a part in making the world a more habitable place for all things.

4 Conclusion and future trends in research

This present volume reveals the importance of considering ethics in agriculture. So often busyness, expectations, and productiveness set the agenda for our lives, our jobs, our industries, and our planet. There is nothing inherently wrong with work, goals, and producing. In fact, such things give our lives meaning and guidance. However, it takes only a little attention to the problems of our world (cultural, agricultural, ecological, etc.) to recognize that our habits, motivations, and projections are leaving our world less tenantable for ourselves and many other living things. In such an environment, we must struggle to find meaningful solutions to these problems we have inherited but also created.

This chapter has suggested that the virtue ethics tradition and Wendell Berry are invaluable resources for addressing many of the agricultural problems facing humans on the earth. But I also have suggested that Berry's virtuous wisdom and knowledge allow for the care of the whole of the material world, not just humans or those few humans who sit atop the pile of earthly spoils. If we take Berry and the virtue ethics tradition seriously for agricultural ethics, I believe at least these three recommendations should guide our agriculture:

1. There must be a reduction in the decision-making and pay gap between those who *perform actual* agricultural work (planting, watering, harvesting, feeding, herding, transporting, slaughtering, etc.) and those who *make decisions about* agricultural work.
2. Good agriculture must view things in a holistic manner. Dan Russell (2013a, p. 2) reveals the importance in the virtue ethics tradition of rejecting urges at compartmentalization, which I suggest resonates with Berry's critiques (1977, p. 41; 1981a, p. xii; 2017a, p. 211) of specialization that do not promote a holism (consideration of all aspects and outcomes) in agriculture.
3. As the existence of this book makes evident, those engaged in agriculture must take an active interest in ethics and what it means for their work. Likewise, those engaged in more formal ethical discourse must find practical connections to real-world matters like agriculture, ecology, labor, and the like.

5 Where to look for further information

I have referenced several times in this chapter the *Cambridge Companion to Virtue Ethics*. Though the first few essays of the volume give good general overviews of the tradition, the volume as a whole is an excellent resource. Additionally, Philip Cafaro and Ronald Sandler have put together two nice edited volumes which explore virtue ethics and environmentalism: *Environmental Virtue Ethics* (2005) and *Virtue Ethics and the Environment* (2010). If one is interested specifically in Aristotle's notion of virtue, Alasdair MacIntyre remains one of the preeminent modern interpreters of the philosopher.

A couple of helpful online (and free!) resources for better understanding virtue ethics include:

- Stanford Encyclopedia of Philosophy (https://plato.stanford.edu/entries/ethics-virtue/).
- Encyclopedia Britannica (https://www.britannica.com/topic/virtue-ethics).

Much of Wendell Berry's work is accessible for a broad readership. I still find his *The Unsettling of America* to be foundational for understanding Berry throughout his career, but it also remains very poignant in our own time. Berry's more recent collections of essays *Our Only World* and *The Art of Loading Brush* are also nice entry points for getting oriented to his thinking. For a good collection of essays on the kind of agrarianism Berry believes in, see Norman Wirzba's edited volume *The Essential Agrarian Reader* (2002).

For examples of agricultural work with virtuousness and sustainability in mind see:

- The Land Institute (https://landinstitute.org/). Founded in 1976 in Salina, Kansas, The Land Institute focuses on 'reconciling the human economy with nature's economy, [believing that] food and how we produce it is a transformative first step. In this future, agriculture regenerates the soil, water, and air upon which all life depends' (https://landinstitute.org/about-us/vision-mission/).
- Shirley Sherrod's New Communities, founded in 1969 in Georgia, is an excellent example of how concerns for racial equity and sustainable agricultural practices can coalesce (https://www.newcommunitiesinc.com/).
- The National Sustainable Agriculture Coalition (NSAC) works in a host of meaningful ways to promote healthy, just, and regenerative agricultural practices. Their website (https://sustainableagriculture.net/) is a great resource for their own work as well as the many other organizations and individuals involved in such agricultural ventures.

6 References

Aristotle. (1962). *Nicomachean Ethics*. Translated by M. Ostwald. Indianapolis, IN: Bobbs-Merrill Educational Publishing.

Berry, W. (1965). The landscaping of hell: Stripe-mine morality in East Kentucky. In: *The Long-Legged House*. Washington, DC: Shoemaker and Hoard, pp. 12-29.

Berry, W. (1977). The ecological crisis as a crisis of agriculture. In: *The Unsettling of America Culture and Agriculture*. New York, NY: Avon Books, pp. 31-42.

Berry, W. (1981a). Forward. In: *The Gift of Good Land: Further Essays Cultural and Agricultural*. San Francisco: North Point Press, pp. ix-xviii.

Berry, W. (1981b). The economics of subsistence. In: *The Gift of Good Land: Further Essays Cultural and Agricultural*. San Francisco, CA: North Point Press, pp. 149-154.

Berry, W. (1981c). Solving for pattern. In: *The Gift of Good Land: Further Essays Cultural and Agricultural*. San Francisco: North Point Press, pp. 134-145.

Berry, W. (2002). Think little. In: Wirzba, N. (Ed.), *The Art of the Commonplace: the Agrarian Essays of Wendell Berry*. Washington, DC: Shoemaker and Hoard, pp. 81-92.

Berry, W. (2012a). It all turns on affection. In: *It All Turns on Affection: The Jefferson Lecture and Other Essays*. Berkeley, CA: Counterpoint Press, pp. 9-40.

Berry, W. (2012b). Landsmen: Jim Leach in conversation with Wendell Berry and Tanya Berry. In: *It All Turns on Affection: The Jefferson Lecture and Other Essays*. Berkeley, CA: Counterpoint Press, pp. 41-66.

Berry, W. (2015). A forest conversation. In: *Our on World: Ten Essays*. Berkeley, CA: Counterpoint Press, pp. 21-52.

Berry, W. (2017a). The order of loving care. In: *The Art of Loading Brush: New Agrarian Writings*. Berkeley, CA: Counterpoint Press, pp. 179-216.

Berry, W. (2017b). The thoughts of limits in a prodigal age. In: *The Art of Loading Brush: New Agrarian Writings*. Berkeley, CA: Counterpoint Press, pp. 19-56.

Bilbro, J. (2019). *Virtues of Renewal, Wendell Berry's Sustainable Forms*. Lexington, KY: University of Kentucky Press, p. 7.

MacIntyre, A. (1981). *After Virtue*. Norte Dame, IN: University of Norte Dame Press.

Plato. The republic. In: Jowett, B. (Trans.), *The Republic and Other Works*. Garden City, NY: Anchor Books, pp. 7-316.

Russell, D. (2013a). Introduction: Virtue ethics in modern moral philosophy. In: *The Cambridge Companion to Virtue Ethics*. Cambridge, UK: Cambridge University Press, pp. 1-6.

Russell, D. (2013b). Virtue ethics, happiness, and the good life. In: *The Cambridge Companion to Virtue Ethics*. Cambridge, UK: Cambridge University Press, pp. 7-28.

Wiebe, J. (2017). *The Place of Imagination: Wendell Berry and the Poetics of Community, Affection, and Identity*. Waco, TX: Baylor University Press.

Zwolinski, M. and Schmidtz, D. (2013). Environmental virtue ethics: what it is and what it needs to be. In: Russell, D. (Ed.), *The Cambridge Companion to Virtue Ethics*. Cambridge, UK: Cambridge University Press, pp. 221-239.

Part 2

Ethical issues

Chapter 7

Agriculture and the environment: ethical issues

Richard Bawden and Roger Packham, Western Sydney University, Australia

For we are conquerors and self-poisoners
More than the scorpion and the snake
And dying of the venoms that we make
Even while you die of us

Judith Wright Australian Poet and environmental activist

1 Introduction

We believe that it is important for us to establish both the context and the perspective of what follows. We need to emphasise that our focus will not be to address what has been recognised as 'a lack of scholarly maturity in the ethics of food and agriculture' (Thompson and Kaplan, 2014) – critically necessary though that certainly is. Such a task is well beyond our competencies, for neither of us are philosophers in the academic sense. Rather, we might best describe ourselves as agricultural scientists and educators with a philosophical

http://dx.doi.org/10.19103/AS.2023.0125.06

bent. Akin to Robert Zimdahl's perspective, we also believe that as agricultural scientists, we have moral responsibilities for nurturing a critical awareness of the importance of agricultural ethics (Zimdahl, 2006). What follows then is offered in the spirit of adding to the critical discourse about ethical issues as they relate to the impact of agriculture on the environment and what that represents with respect to sustainability. This must be discourse which, to borrow from Norton (2005), is rich enough to allow disagreement about value positions as well as allow the integration of scientific knowledge. We come to this as reflections of two agricultural educators who came to ethical consciousness through a somewhat circuitous route which involved the systemic development of strategies for worldview transformation through education for sustainability. Our interpretations of ethics in what follows need to be understood from this perspective. There would be no place on Earth where farming practices have not degraded the environment, directly or indirectly, to some degree or another. Through its contributions to climate change, the indirect environmental impacts of agriculture are now truly global in their influence (IPCC, 2019). The irony is that these impacts now threaten the processes of agricultural production themselves (FAO, 2019). The tragedy is that they also threaten the integrity of the ecology of the entire planet (Karr and Chu, 1995). The paradox is that as human beings we not only know all of this but we recognise its moral implications (Thompson, 1998) - and yet, to all intents and purposes, we fail to heed them. As one commentator has put it, agriculture has 'tumbled us into a mess' from which we are not sure how to escape (Diamond, 1987). The intentional pursuit of the production of 'goods' to meet some of our most basic needs is inevitably associated with some 'bads' (Beck, 1994). Good and bad, along with right and wrong and harm, care, empathy, justice, and equability, are all human values that are associated with moral distinctions. For this reason alone, the impacts that agricultural practices have on the environment critically deserve ethical considerations (Thompson, 1998).

In effect, this has been so since the very dawn of agriculture some 10 000 years or so ago, or perhaps, better stated, 'dawns': it is now clear that agriculture had origins at different times, at different places, under different circumstances, and with the domestication of different plants and animals, all initiatives seemingly independent of each other (Price and Yosef, 2011). In many, if not most cases, these transitions to agriculture from the hunting and gathering that had marked human lifestyles for around 300 000 years (Diamond, 1987) were triggered by changes in the climate. The end of the ice age in the fertile crescent of the Middle East and eventually Europe (Gowdy, 2020) provides a prime example. Whatever the motivation and wherever the location, the adoption of agriculture would prove not only to result in extraordinary changes in the lifestyle of our ancestors but also to changes in the relationships between we humans and the rest of nature - to both 'good' and 'bad' effect. Most importantly, it would also be associated with the emergence of a new belief system or ethos or worldviews among groups of people, also apparently independently of each other. From a prevailing 'nature-centred' (ecocentric) worldview based on humans accepting their status as being 'within nature', the adoption of farming would lead to a shift to a 'human-centred' anthropogenicity where nature is somehow 'out there', to be exploited. Western ethical views, it has been claimed, are essentially anthropocentric (Laal, 2009). This has led to the current situation where we seem to be markedly reluctant to accept that we are in fact an intrinsic species within the natural world which in turn leads to vexing confusion about our actual status as a species (Berkes, 1999). It also contributes to debates about the recognition of values that are intrinsic to nature as revealed, for instance, in discussions around the concept of ecosystem services (Schröter et al., 2014). In this manner, most of those of us who are associated in one way or another with agriculture are caught in a sub-conscious anthropogenic or ecocentric dilemma. This proves incredibly difficult to resolve, in part because, more often than not, we are not aware of this tension or of the nature and significance of our worldviews as determinants of it. In his comprehensive review of the psychology of worldviews, Koltko-Rivera (2004) defines them as sets of personal beliefs and assumptions about the world about us that in large part influence what we each do in it (as well as to it). These include beliefs about the nature of nature and our status as a species with it and reflect what are referred to as our ontologies. A second set of beliefs, our axiologies, relate to the nature of human nature particularly with respect to our values that include ethics and aesthetics. A third set of beliefs refer to what we can call the nature of knowing and knowledge, our epistemologies, which include issues such as what we hold as truth and we can get to know that.

The prevailing values at the 'turn to production' from hunting, gathering, and trapping would have reflected a belief in 'goodness' as an exclusive human benefit. It would come to evolve into the dominant productionist worldview and utilitarian ethic of Western industrialised agriculture (Thompson, 1995). The ends of production would justify the means of getting there as long as the benefits outweighed the costs - including the costs to the environment or to others. The dominating influences of these intellectual and moral stances on how agriculture is practised in all of its multi-dimensions, while seemingly remaining unexamined, certainly deserve to be critically challenged (Zimdahl, 2006). They reflect worldview beliefs and assumptions that are inappropriate and inadequate for current challenges like the environmental impacts of agriculture. Paradigms can be considered to be the action stage of worldviews representing how these latter are expressed in our practices. An intriguing aspect of our individual worldview profiles is that once formed, through a process of development that extends through our youth under significant influences of our experiences in the world about us, they become surprisingly intractable and resistant to challenges for change (Hiebert, 2008). They are also amazingly effective in restricting the search for other alternatives as Norgaard (1994) has noted with respect to paradigms of development. The quest for what has been referred to as an alternative paradigm of agriculture (Beus and Dunlap, 1990), for instance, continues to prove arduous and disappointing in its effectiveness, with instrumental values continuing to prevail over the ethical. There are exceptions, but they are very rare. Charles Massy (2017), who is both an Australian farmer and a scientist, provides one such outstanding example of what a change of worldview of agriculture can yield. With his work on regenerative agriculture, he argues for the integration of values with science, reason with intuition, the modern with the traditional, and theory with practice. Through his efforts in transforming his own sheep farm, within a context of a 'new agriculture: a new earth', he demonstrates what this looks like in practice. So it can be done - but it takes exceptional people to even attempt it!

2 What has happened?

The clearing of land, the deliberate use of fire as a 'management strategy', the mono-cultural conversion of areas of land with the consequential loss of their native biodiversity, the re-direction of waterways for irrigation, and so on, would all have been regarded as 'good' practices at the start. Where there might have been signs of deleterious impacts, they were either cause for shifting to newly cleared land or accepted as necessary 'bads'. The 'good ends' would have been seen as justifying the (sometimes) 'bad means', especially within an overarching spiritual with the instrumental and the intrinsic co-existing as different aspects of the same understanding of reality. As present circumstances so patently illustrate, agriculture would come not only to threaten the harmony of that 'reality' but to do its best to extinguish it. As the centuries passed, the technological would come to usurp the ethical. Environmental damage could always be seen as the incentive for a 'next generation of industrial technologies' rather than as an invitation for critical reflection (Thompson, 1995). The instrumental seems to trump the ethical even under circumstances where some attention has been paid to the latter. For instance, ethical considerations do not rate a mention in a report on the future of food and agriculture issued by the Food and Agriculture Organization (FAO) of the United Nations (FAO, 2007). Yet that organisation has a strong commitment to ethical matters as they relate to agriculture and food (Bhardwaj et al., 2003).

In the process of the industrialisation of agriculture and as a direct consequence of its almost universal diffusion across the globe, great tensions would arise between those embedded within an *ecocentric* ethos and those whose values were assuredly *anthropocentric*. But also in the process, and certainly contributing to the tensions, was the transformation, over time, of relatively benign ecological interventions into ecological assaults and environmental catastrophes. These would appear as environmental degradations and pollutions of the abiotic as well as the biotic elements of the environment, including other humans. The litany of issues that are now recognised as serious in this regard is as long as it is staggeringly diverse and complex (Chua and Karr, 2016). It is of course impossible to establish when these first became evident although it has been claimed that it could be seen from the start, at least in the so-called 'fertile crescent' with deforestation and the erosion and salinisation of the soil (Diamond, 1987). There is no doubt whatsoever that it has reached particular prominence since the early 1950s with the intensification of agriculture particularly with the use of synthetic fertilisers and biocides. This seven-decade-long period has been labelled the 'great acceleration' being marked by excesses of resource use, by the explosion of the human population, and by the deterioration of the entire planetary environment (Steffen et al., 2015). These are clearly all systemically interconnected. It is a sobering fact that in the year 1950 the estimated human population of the world was 2.5 billion which has expanded to now stand at almost 8 billion. It is surely no coincidence that as agricultural production has provided ever more food, so there have been ever more people to consume it. And the more people there are on Earth, the more the environment is degraded.

As climate change is so powerfully illustrating (IPCC, 2019), agriculture, through its very significant contributions to climate change, has impacts on the environment and all associated systemic complexities have truly global consequences. The irony, of course, is that degrading environmental impacts have long been recognised as being detrimental to the aims, purposes, and outcomes of agriculture itself (FAO, 2011), thereby complicating the 'means/ends' narrative of 'production agriculture'. There is little to indicate that it is likely to change, in spite of the eloquence and logic that some outstanding agricultural ethicists continue to bring to the debate. As one of the most pre-eminent of these scholars has argued, even in the face of criticisms, the default response within conventional 'scientific agriculture' is to persist in its faith in the technological rather than to engage with the ethical (Thompson, 2014).

3 Why has it come to this?

With all of our human experiences with agricultural activities over literally millennia and with all of the scientific knowledge and technical know-how that we have acquired over the past couple of 100 years or so, how has this happened? How have agricultural practitioners of so many different hues, done things (continue to do things) that so threaten not just our own well-being but that of all of the rest of nature too? As intelligent beings, why do we continue to foul our own environmental 'nests' without any apparent sense of guilt? Perhaps even more significantly, given that we humans are moral animals (Wright, 1995), *why* have we allowed all of this to happen? Why do we continue to tolerate maleficence in our pursuit of beneficence, to refer to two of the four principles of bioethics (Beauchamp and Childress, 2011)? We all have consciences after all, innate intuitions and feelings that seem to guide us in our judgements about right or wrong that, in turn, urge us into actions (Churchland, 2019). All too often however, a reliance on conscience alone proves too unreliable, too inconsistent, and the source of all-too-frequent conflict within social groupings, to provide solid foundations for sound judgements. Furthermore, we all can exhibit perverse tendencies to ignore what our 'inner voice' seems to be telling us. As human beings, we are, as Wright has so graphically asserted, 'species splendid in our array of moral equipment, tragic in our propensity to misuse it, and pathetic in the constitutional ignorance of the misuse' (Wright, 1995, p. 13).

Fortunately, we are able to move beyond such indeterminacy for we are also all endowed with capacities for ethical consideration (Ayala, 2010). These find expression in our moral identities (Hardy and Carlo, 2011) where morality is conduct that is informed by impartial reason (Rachels, 2003). Moral goodness is the foundation for endowing us with the sense that we are 'worthy human beings' (Pinker, 2008) even if accepting that what is right or feeling what is good does not necessarily lead to doing the right or the good (Gibbs, 2003).

In practical terms, agricultural ethics represents an inquiry into the general question concerning how we *should* act in any given circumstance (Thompson, 2014). With that comes the recognition that we have moral responsibilities to ensure that the 'right thing is done' (Talbert, 2019). If, as has been claimed, every human action has moral significance (Haldane, 2011), then the impacts that agricultural practices have on the environment should certainly be a focus for intense moral consideration. Every human act is, by definition, an intentional and volitional behaviour that has consequences that can be evaluated and adjudged according to estimations of the goodness or badness of its ultimate impacts (Frankena, 1973). So why, in the face of the all-too-familiar list of negative environmental impacts associated with agriculture, do we seem to consistently shirk moral responsibilities? Why, when subjected to the cacophony of criticism from all quarters of society about how agriculture is impacting the environment, do we continue to focus on what *can* be done to mitigate the circumstances rather than what *should* be done to change our entire approach to the issues?

4 Constraints to ethical awareness and moral action

On the face of it, consideration of the environmental impacts of agriculture ought to provide a straightforward entry into the matter of nurturing ethical awareness. Yet to do so is to enter a domain of discomfort, disagreement, and uncertainty where every idea is seemingly contestable once it has been liberated from its shroud of arcane language which, by itself, so inhibits entry in the first place. For all that, as already mentioned, we all do have innate ethical capacities. The ability to evaluate, to possess 'moral sensibilities', results, it has been argued, from our biology which determines three necessary pre-conditions for ethical behaviour: (i) the ability to anticipate the consequences of personal actions, (ii) the ability to make judgements based on human values, and (iii) the ability to choose between alternative courses of action (Ayala, 2010). While these three principles seem clear enough, their application as a guiding practical framework for what has been referred to as 'coming to public judgement' (Yankelovich, 1991) proves to be inadequate. In the first place, having the ability to anticipate the consequences of agricultural activities is not the same thing as actually addressing or seeking to understand their ecological implications. It's not as if there is a lack of awareness of these issues. In a disturbingly comprehensive review, Chua and Karr (2016) catalogue more than one hundred different environmental impact issues that, from our activities as 'ecosystem engineers', threaten to end life on Earth by removing the 'bio' from biosphere. They group them into three categories: (i) indirect depletion of living systems through alterations in physical and chemical environments, (ii) direct depletion of nonhuman life, and (iii) direct degradation of human life. While they represent all of the issues from a generic stance, virtually each one of them is also entirely relevant to agriculture. A cursory glance of reports in the literature not only confirms these observations but reaffirms that the distribution of environmental impacts of agriculture reaches across nations irrespective of development status, culture, political systems, religions, and what might be termed the prevailing ethos or dominant ethic. Groundwater degradation is a serious and persistent hazard worldwide, including in Africa (Xu and Usher, 2006): deforestation in Brazil to clear land for cattle farming and soybean production (to feed cattle in the United States) has reached devastating proportions (da Cruz et al., 2021), as it has in Indonesia to support the creation of palm oil plantations (Schlanger, 2019): soil erosion and land degradation are major problems in India (Anon, 2016) as well as in China, where issues of soil pollution, water scarcity, and loss of biodiversity are all widespread (Cook, 2015). Loss of biodiversity is also of major concern in Europe (Crennat et al., 2019), while in Australia, the Great Barrier Reef is threatened by fertiliser runoff and chemical pollution from agriculture – most especially sugar cane farming (Brodie, 2016). In the Middle East, and especially on the Arabian Peninsula, arable land is being desertified at an alarming rate (Malek, 2018). Soil erosion remains of concern in the United States (UCS, 2020), while traces of the 'forever chemicals' (polyfluorinated alkyl substances (PFAS)) used in food packaging are found in the blood of almost every American (Jaynes, 2022). 'Bads' and wrongs and maleficence galore prevail with even hints of evil – think about the deliberate cultivation of opium poppies in Afghanistan for example or of coca in Colombia or even of tobacco in Cuba for that matter.

As we have long known the scientific and technological foundations of these impacts, why are agricultural practices that lead to them still in vogue? And the same arguments pertain to more recent realisations such as the role that agricultural production and processing practices play in the dispersal of micro plastics, the influence that compounded animal feeds have on the development of antibiotic resistance, and the relationships between land clearance for agriculture and the outbreak of zoonotic diseases like the COVID SARS virus.

How can recourse to ethical considerations help in resolving these issues beyond raising an appreciation that these are indeed moral matters? What practical steps can be taken beyond that and who ought to be involved as agricultural practitioners?

When we talk about actions being taken within agriculture, we are of course talking of a truly enormous constituency. It is estimated that there are some 570 million farms in the world (Lowder et al., 2016). These range from subsistence smallholdings right through to vast international agribusiness corporations with supply chains that are both vertically and horizontally integrated and with a reach that extends across the entire globe. Of note is the fact that, in one form or another, agriculture covers some 50% of all the habitable land on the planet (with 75% of that area devoted to livestock grazing) and uses 70% of water worldwide (FAO, 2011). All up, across the entire global supply chain or ago-ecological system, more than 1 billion people are employed in agriculture (Cassidy and Snyder, 2019) while virtually all of the almost 8 billion of us, rely on agriculture in one way or another. As consumers of food, we all bear some moral responsibility for the way agriculture is conducted (Thomson, 2015). But how do we do that? How do we exercise our moral responsibilities in practice? Perhaps it's already too late. Maybe we have already lost the ability to make ethical judgements as individuals, as a result of two centuries of what Busch (2000) referred to as abdicating our moral responsibility to such 'Leviathans' as the market, the state, and science as well as to the 'church'. But if that is so, then there is an imperative for us to reclaim that which we have lost and to reject the marketism, statism, and scientism to which he refers as adequate systems of belief as foundations for moral responsibility. A focus on 'sustainabilism' instead might prove to be an effective route to recovery (Bawden, 1997).

5 The ability to make judgements

To be effective in the reclamation of our moral competencies, we need to have clarity with respect to what it means to be moral and to act ethically. In essence, we need to address the second of Ayala's principles. Again, it is one thing to claim that we all have the ability to make value judgements. It is quite another to come to an understanding of the nature of human values and their application to judgement, let alone how to put these understandings into actual ethical practice. What is it that makes consequences good, bad or worth being realised in the first place, let alone how do we go about identifying the criteria that should be adopted for the evaluation of consequences (Sinnot-Armstrong, 2019)?

With the establishment of the association of such values as good and bad, right and wrong, and harmful and beneficial, with intentional and volitional acts, we have already identified *consequentialism* as what amounts to one 'school of thought' or ethical theory. There are a number of others that include *rights-based* or *deontological* ethics and *virtue ethics* as two which are also of particular relevance to issues related to the environmental impacts of agricultural practices. Comstock provides a quick entry point into these three formal ethical theories while also identifying the major figures known for their contributions to each one in turn (Comstock, 2002). In the first place, there is *utilitarianism* (with its foundations identified with the nineteenth century Englishmen Jeremy Bentham and John Stuart Mill). This theory is sometimes used interchangeably with *consequentialism* but, as the one who introduced the latter term, Anscombe (1958) made a subtle distinction between them. Where utilitarianism takes account only of the actual consequences of actions, consequentialism determines what should be done solely by reference to the foreseen consequences of actions. Both of these positions, however, concern themselves with the state of difference between good consequences and bad ones arising from agricultural practices. The focus is on the outcomes of the acts themselves. In contrast to this focus, *rights-based* or *duties based, deontological ethics* (identified with the German philosopher Immanuel Kant) call attention to the rights, autonomy, and dignity of others, regardless of consequences. So, where the focus of consequential-based theories is on the acts themselves, deontological ethics focuses on the intentions of those who perform the act. In the third ethical perspective, so-called *virtue ethics* (identified with the contemporary Scot Alasdair MacIntyre and the American Carol Gilligan, but drawing on traditions established by the ancient Greek philosopher Aristotle), the focus is shifted again to embrace the notion that ethical actions should be consistent with ideal human virtues. The emphasis is on patterns of conduct that are considered virtuous with respect to the right, the good and the proper, thus highlighting the importance of the character of the 'actor' as well as the nature of the act (Comstock, 2002).

It is recognised that while each of these different perspectives allow for insights into moral philosophy, they present some significant challenges with respect to ethical practices. In spite of their apparent simplicity and clarity, none of them actually provides a decision procedure that is without some problem or another (Haldane, 2011). Each seems to have significant limitations with respect to establishing exactly what constitutes good and bad, right and wrong, under any particular circumstance, or indicates which particular virtue is relevant (Guinebert, 2020).

As Anscombe observed more than seven decades ago, ethical theory has a tendency to find itself in a quandary marked by the inadequacy of any single approach to meet all circumstances (Anscombe, 1958). Haldane (2011) confirms that the situation has changed little in the interim. Consequentialism, he argues, echoing Anscombe, can't account for all of the complexities and diversities of values nor for the ideal of moral inviolability, while non-consequential approaches remain bedevilled by the sense that the idea of pure practical reason turns out to be either an illusion or to provide, at most, the form but not the substance of ethics (Haldane, 2020). These matters are further complicated by the observation of Bonde and Firenze (2013) that there are further variations within consequentialist, non-consequentialist, and agent-centred theories. We have already noted utilitarianism within the consequentialist approach to matters ethical, but reference is also sometimes made to the egoistic approach and the common ground approach as additional variations on the consequential theme. The non-consequentialist category includes the duty-based approach, the rights approach, the fairness or justice approach, and the divine command approach. And, finally, the virtue approach and the feminist approach are variations on the theme of agent-centred theories (Bonde and Firenze, 2013).

It is not at all surprising that it has been claimed that ethical issues that arise in food and farming defy simple resolution (Mepham, 1996) and that is assuredly so in the face of the extraordinary complexity of the relationships between agriculture writ large and the environments with which they interact. Taken altogether, these moral theories are generally supposed to be mutually exclusive which means that they raise questions about whether or not any particular theory should be accepted as right or true according to particular circumstances (Guinebert, 2020). In principle at least, whenever we face the matter of ethical concerns related to the environmental impacts of agriculture, we should explore the relevance and adequacy of the different approaches in our process of selection (Sinnot-Armstrong, 2019). That of course demands a familiarity with the different theories as well as appropriate competencies in their application in practice particularly as it relates to decision-making in complex and messy situations. Here then is the primary responsibility of us all who are involved in agriculture, especially those of us who are scientists, educators, and policymakers.

6 Applying these concepts in practice

With all of this in mind, is it any wonder that agricultural practitioners privilege the instrumental over the ethical as Thompson (1995) insists? How does the billion-fold constituency get to grasp all of this - or any of it for that matter? There is no doubt that those involved in agriculture would benefit enormously from the sort of reflexivity that Ulrich Beck proposed: for all who are involved in agriculture to engage in critical introspection and self-confrontation with regard to questions related to how we should live our lives. But where is there any encouragement for this to occur? For all the concerns about the impact of climate change on agricultural production and of the impact of the latter on the former, how do we get beyond the discourse of adaptations and mitigations to at least include discussions about our moral obligations, not just to each other but to nature in its entirety? How can we come to accept the intrinsic value of the (rest of the) natural world? What has to happen for us to realise that the degradations that we cause, the polluting agents that we disperse, the excess water that we demand, and the deforestation that we defend all demand critical and urgent ethical attention?

7 Can environmental ethics help?

The domain of environmental ethics potentially has much to offer in this respect, and indeed, Thompson (1995) has claimed that agricultural ethics could sensibly be seen as an aspect of environmental ethics. Yet, to enter that latter world is to open the gates of further esoteric matters, arcane language, and the discomfort of argumentation. While this is the very essence of the work of philosophers, it presents an ambience that is alien to those who are not.

The central thesis of environmental ethics is that the moral concerns of humans for their own good ought not to substitute for moral concerns about nature's own good (Carfaro and Sandler, 2010). In this sense then, environmental ethics is unique because it raises questions about the moral worth of nonhuman and even non-living aspects of the natural world (Rolston, 1975). It demands an ecocentric perspective rather than an anthropocentric one, yet that submission is not without controversy or intellectual challenges (Piccolo, 2010). Ecocentrism assumes that there is objective 'good' in the world that is independent of our human preferences and was surely here long before our species even existed and will assuredly persist long after we have gone (Rolston, 1975). But how is that to be established? Can we transcend the anthropocentric position that it is only we humans that are worthy of ethical consideration (Callicott, 2006)? Given that we are human it would seem that we have no choice but to think like humans even as we might try to grasp the concept of intrinsic value (Crowther, 2019). How can our moral theories be anything but anthropocentric given that they can only be based on what we know or what we care about (Warren, 1997)? Perhaps 'anthropocentrism' is not anthropocentric after all (Kidner, 2014). Perhaps a focus on anthropocentrism distracts us from sensible and necessary discussions about ecocentrism

(Kopninal et al., 2018). Perhaps the distinctions don't matter that much in practice. Isn't anthropocentric motivation the best argument for maintaining the ecological systems upon which we humans rely (Hayward, 1997) and therefore won't that lead to the ultimate convergence on the same practical outcomes as ecocentric positions (Norton, 1984)? All of these issues, remote and esoteric as they might seem, are entirely relevant to the issue of agricultural ethics. But they are clearly not easy to grasp and they are further confounded by a host of other related ideas, concepts, and theories: weak anthropocentrism (Norton, 1984), perspectival anthropocentricism (Ferré, 1996), egocentrism (Rowe, 1990), biocentrism (Berman and Lanza, 2010), anthropoholism (Bassey, 2019a), anthropologism (Bassey, 2019b), biophilia and cosmophilia (Johnston, 2010), human chauvinism and speciesism (Hayward, 1997), industriocentrism (Kidner, 2014), deep ecology (Naess, 1989), feminist ethics (Merchant, 1980), the ecological self (Mathews, 1991), the ethics of place (Smith, 2001), and future ethics (Skrimshire, 2010).

For all of these complications, however, it would seem that these different perspectives represent variations on the foundational theme of environmental ethics as it was first conceived and presented by two different scholars working in two different countries in the mid-1970s. Both of their seminal pieces were posed as responses to similar self-set questions. Richard Routley (Sylvan) in Australia presented persuasive arguments in response to his question *Is There a Need for a New, an Environmental Ethic?* (Routley, 1973). Holmes Rolston III an American philosopher addressed the not-dissimilar question *Is There an Ecological Ethic?* (Rolston, 1975). Of particular interest in the present context is that both philosophers had drawn upon and critically addressed the insights of Aldo Leopold related to the *land ethic* which he had proposed in his book *The Sand County Almanac* (Leopold, 1949). Leopold, who was not a philosopher but a scientist, forester, conservationist, and hunter, is also widely known beyond philosophy for his maxim that:

> a thing is right when it tends to preserve the integrity, stability, and beauty of the biotic community. It is wrong when it tends otherwise. (pp. 224–225)

This statement, as a synopsis of his position on a land ethic, has been subjected to a wide variety of criticisms from some, while drawing strong support from others. Do biotic communities or ecosystems possess the properties of integrity and stability? Do such communities or ecosystems even 'exist in reality' or are they merely mental constructs? If they are indeed the way the natural world is organised, where do *Homo sapiens* fit? And as far as beauty is concerned, does that not just lie in the eye of the beholder? How holistic was Leopold in his approach to nature? As Eba (2020) sees it, he was not 'holistic enough' while the debate continues about whether the land ethic represents an ecocentric perspective (Callicott, 2011) or an anthropocentric one (Comstock, 1994). Budolfson (2014) has suggested that much of the controversy results from a misinterpretation of Leopold's position and especially through the isolation of the two sentences of the 'maxim' from the rest of the short land ethic section of the book. The sentence preceding the 'maxim' statement provides an all-important context: 'examine each question in terms of what is ethically and aesthetically right, *as well as what is economically expedient*' (emphasis added). In other words, rather than presenting a prescription of what an ethical approach to land management should look like, Leopold raised issues to do with the necessary associations between ecology, evolution, economics, and ethics. Vitally, he came to these understandings and insights in large part through reflecting on his own experiences of the natural world: his own experiential participation in the wholeness of nature, arguing that we can only be ethical 'in relation to something that we can see, feel, understand, love, or otherwise have faith in' (Leopold, 1949, p. 214). Nelson (1998) emphasises Leopold's point that the land ethic might be considered 'an attempt to change people's rudimentary thoughts about the land'. Leopold was mostly 'in the business of worldview transformation; indeed, worldview revolution' (Callicott, 2011, p. 511).

8 Ethics and worldview transformation

We cite Leopold here for, upon reflection, we believe that we can situate our own initiatives in agricultural education in Australia within the spirit of Leopold's land ethic. While regrettably unaware of his work when we started our radical initiatives in curriculum reform in the late 1970s at Hawkesbury Agricultural College (Bawden et al., 1984), we too would eventually find ourselves, as educators, in 'the business of worldview transformation'. We would come to recognise the significance of facilitating learning through critical reflections on experiences in the 'real world' of everyday agriculture. And we would help our students grasp the significance of 'holism' through a focus on their development of what we came to portray, following Salner (1986), as systemic competencies. We would have appreciated Leopold's position where science meets philosophy in such a way that together they present a profound challenge to the prevailing productionist or mechanistic worldview of agriculture. We most certainly would have grasped the way he derived his thoughts and feelings and ideas through reflection on his own lived experiences as a forester, a conservationist, and a hunter. While it was certainly not the original motivation for our endeavours, we would come to contribute to an extended and continuing critical discourse about the clear inadequacies of that dominant worldview, which we would present as technocentric. In our quest as a learning community to understand sustainability as it related to the development of agriculture, we would come to focus explicitly on worldview awareness and transformation.

We can claim that we too developed something resembling a maxim:

The responsible transformation of 'agricultural systems' in the 'real world' demands and depends on the transformation of the intellectual and moral systems that constitute our worldviews.

We are making an important distinction here that many of those who think about, research into, and comment upon environmental as well as agricultural ethics seem to miss but which Leopold certainly grasped: if we want to change the way we treat the world, we need to change the way we 'see' it. We need to critically re-assess the way we experience it, feel it, perceive it, make sense out of it, and challenge our foundational worldview beliefs and attitudes. The transformation of our worldviews is a pre-requisite for transforming the world. It's not just that the resort to the instrumental represses opportunities for ethical reflection as Thompson (1995) has submitted, but it effectively suppresses any critical reflection at all. A basic problem is that the vast majority of us don't seem to appreciate that we even have worldviews let alone what constitutes them, and without such fundamental awareness, any attempt at transformation can rest on little more than 'blind persuasion'. As humans, we remain fixated on the ecosystems, agri-food systems, agro-ecological systems, farming systems, and so on 'out there' without a thought about the mental systems that we use to generate, attribute properties to, and attempt to engineer these constructs. It was relatively early in the development of our own initiatives as agricultural educators at Hawkesbury that we recognised the profound importance of the 'shift in systemicity' from investigating the 'systems out there' to particular ways of thinking and acting systemically with respect to those issues (Bawden and Packham, 1993).

9 Worldviews, experiential learning and systemic development

A vital aspect of our work has been the attention we have paid to the nature of the worldview transformations that we ourselves have experienced as agricultural scientists. Of signal importance in this regard has been the significance of the synthesis of the innately dynamic experiential learning processes with what we refer to as systemic perspectives. Following Kolb (1984), we came to understand experiential learning as a rigorous reflective, recurrent, holistic, and dialectical process of 'making sense' out of what is being experienced and 'taking action' appropriate to the circumstances. The 'making sense' phase draws on concepts, principles, and theories as abstractions, while the 'taking action' stage demands relevant practices to accommodate change.

The process assumes a reflexive dimension as learners learn how to confront the impact that their own worldview beliefs and assumptions are having on all aspects of the process. The adoption of systemic perspectives is a critical focus of this reflexivity. We have come to believe that this synthesis is a critical foundation in the quest for sustainability within an agricultural context, in all of the complexities, uncertainties, and ambiguities that are associated with that concept. Vital insights from educational theory, informed by the work of Perry (1968) in particular, led us to understand the concept of different stages of intellectual and moral development. From the work of Kitchener (1983) and Salner (1986) on epistemic development, we came to appreciate the significance as well as the challenges of facilitating the intellectual and moral maturation crucial for addressing the implications and applications of systems thinking. Simply put, the matter of sustainability is of such complexity that it demands refined systemic appreciation, and thus systems thinking capacities that involve both intellectual and moral development, that in turn form the foundations for the transformation of worldviews.

A concern for the sustainability of Australian agriculture was a central entry point into the work that we would eventually describe as systemic development (Bawden, 2005). The context of our initiatives was to respond to two challenges. The first was an internally generated mandate from the college board to reform the curriculum from its technical focus to one that was 'more scientific and modern'. Yet we were hesitant about such a direction due to the second challenge, our sensitivity to a virtual cacophony of complaints about the environmental impacts of agriculture that were coming from all quarters. These voices included environmentalists, conservationists, consumers, academics, journalists, research scientists, politicians, and even poets. And farmers themselves, conscious of all of this, were also all too aware that they were caught in a vicious downward spiral of declining terms of trade or the cost-price squeeze. They had to keep increasing their productivity just to keep pace with this difference between the costs of their inputs and the returns from the commodities that they produced. Reparations were expensive and seemingly impermanent.

It was not difficult to conclude that Australian agriculture was not sustainable on environmental grounds alone, to say nothing of its undesirable impacts on rural communities and other domains within Australian society. And we were certainly not convinced that increasing the science and economics content of our curriculum would be of much help under the circumstances. This is in spite of the reassurances that we kept hearing from our 'mainstream' agricultural scientists, economists, and academic colleagues that, as in the past, the prevailing challenges would be met essentially through the development, diffusion, and adoption of innovative technologies.

10 A worldview of persistence

The emerging concerns for sustainability from within the discourse on agricultural development presented an opportunity for us to confront the challenges that a worldview for persistence would present. From the start, we recognised that would present an arena of investigation that would not be without difficulty. As Douglas (1984) would note, sustainability means different things to different people, and while he did not refer to it specifically, the subtext was that each of us as individuals, as well as members of particular social collectives, hold a range of different worldview beliefs and assumptions. It is interesting to note that such diversity continues within the sustainability discourse even extending to the matter of interpretations of the ethics of sustainability (Raffalle et al., 2010).

The challenge for our Hawkesbury team was how to approach education for sustainability with respect to Australian agriculture given the confusion and ambiguities not just of interpretations of the meaning of the construct but also of its conceptual foundations. One way forward presented itself through the publication of attempts to bring order to the chaos of sustainability in agriculture by Douglas (1984). He distinguished between what he referred to as 'three schools' of approach to the matter: the 'food sufficiency' or 'productivity' school, the 'stewardship' school, and the 'community' school. After describing what he meant by each of these different views of sustainability (effectively worldviews), he came to the sombre conclusion that it was the production focus of the food sufficiency view that not only dominated agriculture but showed little evidence of vulnerability to challenge by the others (Douglas, 1984). Paul Thompson, in his seminal *Spirit of the Soil*, further developed these positions recognising four different 'worldviews' (his designation) as philosophical approaches to agriculture with particular emphasis on its relationships with the environment (Thompson, 1995). The four perspectives that he presented, and which he explored in considerable detail in his text, he referred to as productionism, economy, stewardship, and holism. Echoing Douglas, he also concluded that productionism, with its emphasis on the primacy of yield as the primary focus for adjudging value, was the dominant worldview in practice. And this, he asserted, was both supported and endorsed by all of the ancillary institutions of the 'agricultural establishment' including scientific research and education with their positivist and reductionist intellectual foundations. By coincidence, we had also identified four different worldviews related to the challenges of sustainability in agriculture, on this occasion within an Australian context which we labelled egocentric, technocentric, egocentric, and holocentric. In contrast to Thompson, who presented his different worldviews as alternatives to each other, we saw ours as representing a developmental sequence of worldviews in the order presented above. Following insights drawn from publications in sociology

(Cotgrove, 1982), environmental psychology (Miller, 1985), naturalistic research (Lincoln and Guba, 1985), and organisational sociology (Burrell and Morgan, 1979), we used two very simplified binary distinctions to create an orthogonal matrix using ontological (nature of nature) and epistemological (nature of knowledge) distinctions as the 'parameters'. The ontological distinctions were between holism (H) and reductionism (R) (atomism) as polar opposites on the vertical axis and objectivism (O) and contextualism (C) as opposing perspectives on knowledge and ways of knowing on the horizontal axis. We identified the four cells of the resulting matrix as representing egocentric (RC), technocentric (RO), ecocentric (HO), and holocentric (HC) worldviews, respectively. Using this framework, we were able to attribute four different worldview conceptions of sustainability which overlapped to some degree with some of the positions adopted by Douglas and Thompson, particularly with respect to productionism and technocentrism. This conceptual framework was then adopted as a curriculum guide for experiential and systemic pedagogical practices that were designed to expose our students (and later consultancy clients) to the nature and significance of worldviews in agricultural development. In a radical departure from the conventions of agricultural education, our students would learn about the different perspectives and the difficulties involved in adopting them through reflections on their own experiences in 'real-world' situations that were designed to increase in complexity and messiness over the duration of the program. As the course progressed, the students would be almost literally provoked to reflexively confront their own beliefs and value assumptions and assess their adequacy in the face of the search for the sustainable, systemic development of Australian agriculture into the future. While there was no explicit inclusion of the axiological (value-based) dimension in our worldview matrix (three dimensions would prove very difficult to use as a heuristic), we surmised, correctly as it turned out, that ethical positions would become evident under interrogation of the other two. Thus, ontological questions about how one 'sees' one's position in nature almost certainly raise issues related to intrinsic values while considerations of the ecological roles that one plays (as well as humans collectively) also lead to an exposure of value judgements related to notions of damage and harm and 'betterment' and so on. Epistemological interrogations were further related to what one knows about anything (especially about self) as well as how one comes to know that. It is impossible to raise ethical and moral consciousness without knowing the nature of knowledge as well as how it can be known. That stated, our Hawkesbury initiatives would most certainly have benefited from the insights from formal moral and ethical dimensions of agriculture and particularly as they pertained to the systemic development of agriculture in Australia. But the publication of *Spirit of the Soil* came too late for us. In the very year of the release of that wonderful book in 1994, the new administration of our institution commenced processes that would eventually lead to the abandonment of the initiatives (Bawden, 2005).

The positivist and reductionists had won and productionism would again be restored as the ethos prevailing in the remnants of agricultural education at the institution which is now a part of Western Sydney University.

While it is certainly true that currently there are other significant initiatives related to sustainability and sustainable development within this university, none appear to be addressing the issue from a worldview transformational perspective. And the same can be said, with very few exceptions, for what is happening within Australian agriculture writ large. The supreme irony is that the aboriginal peoples who have lived on this land for as long as 60 000 years have a very profound sense of the significance of their 'worldview' beliefs and assumptions to the way that they live(d) their lives. And while they certainly don't use the word worldview, there are certainly indications that ontologies, epistemologies, and axiologies can be interpreted from some of their accounts. They have been holocentric all along!

11 An Australian aboriginal perspective

We close this chapter with a short commentary in this last regard from Bob Randall who is an elder of the Anangan peoples of Central Australia which was recorded in a 2006 documentary (Hogan, 2006), and which here draws on material that we have previously published (Bawden, 2010). Randall argues that the overall commitment of his people is to 'responsible connectedness' - *Kanyini* in the Anangan language. The foundations for this commitment, *Tjukurrpa*, provide the 'overarching system of beliefs' that represent what is referred to (somewhat misleadingly) in English as the *Dreaming* (Price-Williams and Gaines, 1994). As the Encyclopedia Britannica describes it, the *Dreaming* or dream-time, or world dawn, is 'a mythological period of time that had a beginning but no foreseeable end during which the natural environment was shaped and humanized by the action of mythic beings'. *Tjukurrpa*, in turn, comprises three elements or 'pillars' each of which can be interpreted as having a rough equivalence to ontological, epistemological, or axiological beliefs and assumptions. *Ngura* is the belief that land is both nature and an all-encompassing home. This represents an ontology of integration that stands in direct opposition to the 'Western' notion of land as just land, but not dissimilar to Leopold's land ethic. From this perspective, land cannot be owned or reduced to property, and under no circumstances should it be fouled. *Kurunpa* introduces the profound pantheistic spirituality that includes respecting the sanctity of the ancestors, which presents a metaphysical cosmological perspective to *Tjukurrpa*. Finally, *Walytja* represents family in the broadest possible sense of 'responsible connectedness' not just among humans but vitally between humans and the land. This perspective can be interpreted as representing axiological concerns through its emphasis on justice and equitability and responsibilities for the care of 'the other' which is actually not the other at all, but an intrinsic part of the 'family'.

Randall's commentary is not merely an account of what might be described as the worldview foundations of the Anangan peoples of Central Australia but a searing indictment of the assaults that the late eighteenth-century British colonists (and their successors) inflicted on all aspects of *Kanyini*. He takes each of the elements in turn and describes in detail, how particular activities, associated especially with 'invasive agriculture', not only offended but, to a great extent, attempted to virtually extinguish the environmental ethic of an entire culture. It is not as if these indigenous peoples had had no significant impacts on the environment during their sustained presence on the land over very many millennia. Through their controlled but very extensive use of fire as a technical aid to hunting, aboriginals had undoubtedly altered ecological relationships, although there is considerable debate about the actual extent and character of these changes (Bowman, 2008). There is certainly little doubt that they changed the landscape through their burning practice (Gammage, 2011). They also contributed to the extinction of the megafauna that had once roamed the land (Choquenot and Bowman, 1998). There is also increasing evidence that, contrary to long-held beliefs by non aboriginal people, the indigenous Australians developed some agricultural practices which were, however, regarded as complementary activities to hunting, gathering, and trapping rather than as lifestyle occupations (Gerritson, 2008).

Essentially however, the arrival of the colonists and their introduction almost literally overnight, of a host of species of crop seeds, livestock animals, and other plants and animals that were truly exotic, represented nothing less than an 'invasion ecology' (Elton, 1958) as well as an invasion 'social ecology' (Bawden, 2010): both on a massive scale. Neither the land nor the belief systems of the indigenous peoples would ever be the same again.

12 Conclusion

There is obviously no call to attempt to adopt any of these actual practices to amend the ways of industrial agriculture in response to its continuing deleterious impacts on the environment, including the atmosphere. There could be much to gain, however, by turning our attention to the wisdom inherent in what can be interpreted as similar philosophical foundations presented by both Leopold's land ethic and Randall's Anangan people's *Kanyani*. In both cases, ethics, as axiologies, are but one dimension of the set of beliefs and assumptions that comprise our individual and cultural worldviews. From this perspective, it makes little sense to focus on them exclusively without equal consideration of the ontologies and epistemologies with which they are so systemically bound. It is of considerable interest to note that this concept of worldviews as integrated 'systems of meaning and meaning making' (Hedlund-de Witt, 2013) is gaining prominence in the environmental sustainability discourse (Hulme, 2009). The call is for 'an integral ecology of worldviews as a framework for their necessary transformation within the context of climate change' (de Witt and Hedlund, 2017). Consideration of our beliefs and assumptions related to human values like ethics and aesthetics in isolation from those related to those of the nature of nature and the nature of knowledge runs counter to the holistic ideal. The emergence of holistic meaning and the motivation for systemic action are dulled. One of the key principles of systems theory is that whole entities (real or abstract) have emergent properties. Worldviews, as onto-axio-epistemological systems, are no exception. It is thus the transformation of the whole positivist-reductionist-instrumental worldview that is indicated if the dominance of the productionist-technocentric 'world view' is to be effectively usurped. Perhaps this lack of integration is contributing to the submission that awareness of ethics and even scholarly knowledge of moral philosophy, for all its scholarship, has yet to mount a major challenge to the dominant ethic of agriculture (Zimdahl, 2006). We believe that our own initiatives in the facilitation of worldview transformation, based on the principles of facilitated intellectual and moral development, have illustrated the practical potential in this regard.

As 're-born agricultural scientists', we might best describe ourselves as Goethian. Writing of the science of the nineteenth polymath Johann Wolfgang von Goethe, Bortoft (1999) presents the following: 'the Goethian scientist does not lose himself or herself in nature, but finds nature *within* himself/herself in fully conscious experience .... It is a synergistic condition in which humanity and nature work together in such a way that each becomes more fully itself through the other. Both are enhanced but only within their working together, because there is one occurrence which is the mutual enhancement of *both*' (emphasis in the original) (Bortoft, 1996, p. 271).

13 Where to look for further information

The following four books are very accessible introductions to agricultural ethics: *Agriculture's Ethical Horizons* 2006 *by* Robert Zimdahl published by Elsevier Press, NY; *From Field to Fork* 2015 by Paul Thompson published by Oxford University Press; *The Spirit of the Spoil* 2017 (Second Edition) published by Taylor and Francis; *Agricultural Ethics; Research, Teaching and Public Policy* 1998 by Paul Thompson published by Iowa State University Press.

The *Encyclopaedia of Food and Agricultural Ethics* 2014 edited by Paul Thompson and David Kaplan and published by Springer Reference provides a comprehensive coverage of the field.

Relevant journals include *Journal of Agricultural and Environmental Ethics, International Journal of Sustainable Agriculture, and Agriculture and Human Values.*

The collected papers edited by Jules Pretty in Earthscan's *Sustainable Agriculture and Food* 2007 is still among the most comprehensive coverage of the central issues related to this topic.

Critical Systems Thinking and the Management of Complexity 2019 edited by Michael Jackson published by Wiley; *Systemic Intervention* 2001 edited by Gerald Midgley published by Springer are the two books that are invaluable introductions to systems thinking and practices.

Worldview: The History of a Concept 2002 by David Naugle is a valuable resource for those seeking to understand the nature and significance of worldviews.

14 References

Anon (2016). Status of soil erosion or land degradation in India December 2016 conference: managing soil resources for environmental sustainability: Challenges & perspectives. DOI: 10.13140/RG.2.2.33327.59040 (Accessed February 21, 2022).

Anscombe, G. E. M. (1958). Modern moral philosophy. *Philosophy* 33(124): 1–19.

Ayala, F. J. (2010). Colloquium paper: The difference of being human: Morality. *Proceedings of the National Academy of Sciences of the United States of America* 107(Suppl. 2): 9015–9022. DOI: 10.1073/pnas.0914616107 (Accessed February 8, 2022).

Bassey, S. A. (2019a). 'Anthropoholism' as an authentic tool for environmental management. *International Journal of Environmental Pollution and Environmental Modelling* 2: 160–168. Available at: https://philpapers.org/rec/BASAAA-2.

Bassey, S. A. (2019b). Anthropogolism. *Journal of V. N. Karazin Kharkiv National University, Series "Philosophy: Philosophical Peripeteias"* 60. DOI: 10.26565/2226-0994-2019-60-8 (Accessed April 18, 2022).

Bawden, R. J. (1997). Learning to persist: A systemic view of development. In: Stowell, F. A., Ison, R. L., Armson, J., Holloway, S. J. and McRobb, S. (Eds.), *Systems for Sustainability: People, Organizations and Environments. Proceedings of the Annual Conference of the UK Society for Systems Sciences*. Milton Keynes. Plenum Press, New York. UK.

Bawden, R. J. (2005). Systemic development at Hawkesbury: Some personal lessons from experience. *Systems Research and Behavioral Science* 22(2): 151–164.

Bawden, R. J. (2010). Epistemic aspects of social ecological conflict. In: Wright, D., Camden-Pratt, C. and Hill, S. (Eds.), *Social Ecology: Applying Ecological Understanding to Our Lives and Our Planet*. Hawthorn Press, Stroud, pp. 52–63.

Bawden, R. J., Macadam, R. D., Packham, R. J. and Valentine, I. (1984). Systems thinking and practices in the education of agriculturalists. *Agricultural Systems* 13(4): 205–225.

Bawden, R. J. and Packham, R. G. (1993). Systemic praxis in the education of the agricultural practitioner. *Systems Practice* 6(1): 7–19.

Beauchamp, T. L. and Childress, J. F. (2011). *Principles of Biomedical Ethics*. Oxford University Press, New York.

Beck, U. (1994). Chapter 1: Reflexive modernisation: Politics, tradition and aesthetics in the modern social order. In: Beck, U., Giddens, A. and Lash, S. (Eds.), *The Re-Invention of Politics: Towards a Theory of Reflexive Modernization*. Stanford University Press, Stanford, pp. 3–55.

Berkes, F. (1999). *Sacred Ecology: Traditional Ecological Knowledge and Resource Management*. Taylor & Francis, Philadelphia.

Berman, B. and Lanza, R. (2010). *How Life and Consciousness Are the Keys to Understanding the True Nature of the Universe*. BenBella Books, Dallas. TX.

Beus, C. E. and Dunlap, R. E. (1990). Conventional versus alternative agriculture: The paradigmatic roots of the debate. *Rural Sociology* 55(4): 590–616.

Bhardwaj, M., Maekawa, F., Niimura, Y. and Macer, D. R. J. (2003). Ethics in food and agriculture: Views from FAO. *International Journal of Food Science and Technology* 38(5): 565–577.

Bonde, S. and Firenze, P. (2013). *A Framework for Making Ethical Decisions*. https://brown.edu/academics/science-and-technology-studies/framework-making-ethical-decisions (Accessed April 3, 2022).

Bortoft, H. (1996). *The Wholeness of Nature: Goethe's Way Toward a Science of Conscious Participation in Nature*. Lindisfarne Books, New York.

Bowman, D. M. J. S. (1998). The impact of Aboriginal landscape burning on the Australian biota. *New Phytologist* 140(3): 385–410. DOI: 10.1111/j.1469-8137.1998.00289.x. https://www.researchgate.net/publication/229720620_The_impact_of_Aboriginal_landscape_burning_on_the_Australian_biota (Accessed April 4, 2022).

Brodie, J. E. (2016). *Terrestrial pollutant run-off to the Great Barrier Reef: Effects, causes, sources, and management*. PhD Thesis, James Cook University. Available at: http://research.jcu.edu.au/48864/ (Accessed May 19, 2022).

Budolfson, M. B. (2014). Why the standard interpretation of Aldo Leopold's land ethic is mistaken. *Environmental Ethics* 36(4): 443–453.

Burrell, G. and Morgan, G. (1979). *Sociological Paradigms and Organisational Analysis*. Heinemann Educational Books, New York.

Busch, L. M. (2000). *The Eclipse of Morality*. Aldine de Gruyter, New York.

Callicott, J. B. (2006). Conservation values and ethics. In: Groom, M. J., Meffe, G. K. and Carroll, C. R. (Eds.), *Principles of Conservation Biology*. Sinauer, Sunderland.

Callicott, J. B. (2011). *The Worldview Concept and Aldo Leopold's Project of 'World View' Remediation*. Equinox Publishing Ltd., Sheffield. DOI: 10.1558/jsrnc.5i4.510.

Carfaro, P. and Sandler, R. D. (2010). *Virtue Ethics and the Environment*. Springer, New York.

Cassidy, E. and Snyder, A. (2019). *How Many People Work in Agriculture? Resource Watch*. Available at: https://blog.resourcewatch.org (Accessed March 11, 2022). Map of the Month.

Choquenot, D. and Bowman, D. M. J. S. (1998). Marsupial megafauna, aborigines and the overkill hypothesis: Application of predator-prey models to the question of Pleistocene extinction in Australia. *Global Ecology and Biogeography Letters* 7(3): 167–180.

Chua, E. W. and Karr, J. R. (2016). *Environmental Impact: Concept, Consequences, Measurement*. Published Online October 31, 2016. DOI: 10.1016/B978-0-12-809633-8.02380-3 (Accessed March 18, 2022).

Churchland, P. (2019). *Conscience: The Origins of Moral Intuition*. W.W Norton, New York.
Comstock, G. L. (1994). Do agriculturalists need a new, and ecocentric, ethic? Presidential address to the agriculture, food, and human values society. *Agriculture and Human Values* 12: 2–16.
Comstock, G. L. (2002). Ethics. In: Comstock, G. L. (Ed.), *Life Science Ethics*. Iowa State Press, Ames.
Cook, S. (2015). IIED sustainable Ag in China then and now. Seth Cook blog March 26, 2015. IIED Sustainable Ag in China then and now. *International Institute for Environment and Development*. Available at: https://www.iied.org/sustainable-agriculture-china-then-and-ow#:~:text=Chinese%20agriculture%20currently%20faces%20major,of%20agricultural%20biodiversity%20are%20widespread (Accessed May 15, 2022).
Cotgrove, S. (1982). *Catastrophe or Cornucopia: The Environment, Politics and the Future*. John Wiley and Sons, Chichester.
Crennat, E., Sinkkos, T. and Sala, S. (2019). Biodiversity impacts due to food consumption in Europe. *Journal of Cleaner Production* 227: 378–391. DOI: 10.1016/j.jclepro.2019.04.054 (Accessed April 4, 2021).
Crowther, R. (2019). *Well-Being and Self-Transformation in Natural Landscapes*. Palgrave Macmillan, London.
da Cruz, D. C., Benayas, J. M. R., Ferreira, G. C., Santos, S. R. and Schwartz, G. (2021). An overview of forest loss and restoration in the Brazilian amazon. *New Forests* 52(1): 1–16. DOI: 10.1007/s11056-020-09777-3 (Accessed April 5, 2022).
De Witt, A. and Hedlund, N. (2017). Towards an integral ecology of worldviews: Reflexive communicative action for climate solutions. In: Mickey, S., Kelly, S. and Robbert, A. (Eds.), *The Variety of Integral Ecologies*. SUNY Press, New York, pp. 305–344.
Diamond, J. (1987). The worst mistake in the history of the human race. *Discover Magazine*. Available at: https://www.agron.iastate.edu/courses/qgron342/diamondmistake.html (Accessed March 15, 2022).
Douglas, G. K. (1984). The meanings of agricultural sustainability. In: Douglas, G. K. (Ed.), *Agricultural Sustainability in a Changing World Order*. Westview Press, Boulder.
Eba, M.-A. (2020). A critique of Aldo Leopold land ethic for environmental management. *Journal of Office* 6(July–December). DOI: 10.26858/jo.v612.17974 (Accessed April 5, 2022).
Elton, C. (1958). *The Ecology of Invasions by Animals and Plants*. Methuen, London.
FAO (2007). *The Future of Food and Agriculture: Trends and Challenges*. Food and Agriculture Organization, Rome. Available at: https://www.fao.org/3/i6583e/i6583e.pdf (Accessed March 5, 2022).
FAO (2011). The state of the World's land and water resources for food and agriculture: managing systems at risk. *Earth Scan NY*. Available at: https://www.fao.org/3/11688e/i1688e.pdf (Accessed March 5, 2022).
FAO (2019). FAO's work on climate change. Available at: https://www.fao.org/3/ca7126en/ca7126en.pdf (Accessed April 20, 2022). United Nations Climate Change Conference 2019.
Ferré, F. (1996). Persons in nature: Towards an applicable and unified environmental ethics. *Ethics and the Environment* 1: 15–25.
Frankena, W. F. (1973). *Ethics*. Prentice-Hall, Englewood Cliffs, NJ.
Gammage, B. (2011). *The Biggest Estate on Earth: How Aborigines Made Australia*. Allen and Unwin, London.

Gerritson, R. (2008). *Australia and the Origins of Agriculture*. BAR International Series 1874. Archaeopress, Oxford.

Gibbs, J. C. (2003). *Moral Development and Reality: Beyond the Theories of Kohlberg and Hoffman*. Sage Publications, London.

Gowdy, J. (2020). Are hunter-gatherers the future: Climate change, agriculture and un-civilisation. *Futures* 115: 102488. DOI: 10.1016/j.futures.2019.102488 (Accessed March 3, 2022).

Guinebert, S. (2020). How do moral theories stand to each other? Some moral metatheoretical thoughts. *Zeitschrift fur Ethik und Moralphilosophie* 3(2): 279–299.

Haldane, J. (2011). Is every action morally significant? *Philosophy* 86(3): 375–404. DOI: 10.1017/S0031819111000192 (Accessed March 3, 2022).

Hardy, S. M. and Carlo, G. (2011). Moral identity: What is it, how does it develop, and is it linked to moral action? July 2011. *Child Development Perspectives* 5(3): 212–218. DOI: 10.1111/j.1750-8606.2011.00189.x (Accessed March 3, 2022).

Hayward, T. (1997). Anthropocentrism: A misunderstood problem. *Environmental Values* 6(1): 49–63.

Hedlund-de Witt, A. (2013). Worldview is and their significance for the global sustainable development debate. *Environmental Ethics* 35(2): 133–162.

Hiebert, P. G. (2008). *Transforming Worldviews: An Anthropological Understanding of How People Change*. Baker Academic, Grand Raids.

Hogan, M. (2006). (producer) Kanyini. Available at: http://wwwkanyini.com.

Hulme, M. (2009). *Why We Disagree About Climate Change: Understanding Controversy, Inaction and Opportunity*. Cambridge University Press, Cambridge.

IPCC (2019). *Climate Change and Land: An IPCC Special Report on Climate Change, Desertification, Land Degradation, Sustainable Land Management, Food Security, and Greenhouse Gas Fluxes in Terrestrial Ecosystems*. Available at: www.IPCC.ch (Accessed March 15, 2021).

Jaynes, C. H. (2022). Scientists say it's time to phase out plastics to stop sea of pollution. *Ecowatch*, May 9. Available at: https://www.ecowatch.com/scientst-plastic-production-phaseout.html (Accessed April 23, 2022).

Johnston, L. F. (2010). From biophilia to Cosmophilia: The role of biological and physical sciences in promoting sustainability. *Journal for the Study of Religion, Nature and Culture* 4(1). DOI: 10.1558/jsrnc.v4i1.7 (Accessed April 6, 2022).

Karr, J. R. and Chu, E. W. (1995). Ecological integrity: Claiming lost connections. In: Westra, L. and Lemons, J. (Eds.), *Perspectives in Ecological Integrity*. Kluwer Academic Publishers, Dordrecht, pp. 34–48.

Kidner, D. W. (2014). Why anthropocentrism is 'not anthropocentric'. *Dialectical Anthropology* 38(4): 465–480.

Kitchener, K. S. (1983). Cognition, metacognition, and epistemic cognition. *Human Development* 26(4): 222–232.

Kolb, D. A. (1984). *Experiential Learning: Experience as the Source of Learning and Development*. Prentice Hall, New Jersey.

Koltko-Rivera, M. E. (2004). The psychology of worldviews. *Review of General Psychology* 8(1): 3–58.

Kopninal, H., Wasington, H., Taylor, B. and Piccolo, J. J. (2018). Anthropocentrism: More than just a misunderstood problem. *Journal of Agricultural and Environmental Ethics* 31: 109–127.

Laal, M. (2009). A brief history of Enviroethics and its challenges. *Journal of Medical Ethics and History of Medicine* 2: 10.
Leopold, A. (1949). *A Sand County Almanac: And Sketches Here and There*. Oxford University Press, Oxford.
Lincoln, Y. S. and Guba, E. G. (1985). *Naturalistic Inquiry*. Sage Publications, Beverly Hills, CA.
Lowder, S. K., Skoet, J. and Raney, T. (2016). The number, size and distribution of farms, smallholder farms, and family found worldwide. *World Development* 87: 16–29.
Malek, C. (2018). 17 October 2018. *Arab News*.
Massy, C. (2017). *Call of the Reed Warbler: A New Agriculture, a New Earth*. University of Queensland Press, Brisbane, Australia.
Mathews, F. (1991). *The Ecological Self*. Routledge, New York.
Mepham, B. (1996). *Bioethics for the Biosciences: An Introduction*. Oxford University Press, Oxford.
Merchant, C. (1980). *The Death of Nature: Women, Ecology and the Scientific Revolution*. Harper & Row Publishers, San Francisco.
Miller, A. (1985). Cognitive styles and environmental problem-solving. *International Journal of Environmental Studies* 26(1–2): 21–31.
Naess, A. (1989). *Ecology, Community and Lifestyle: Outline of an Ecosophy* (Translated by D. Rothenberg). Cambridge University Press, Cambridge.
Nelson, M. P. (1998). Aldo Leopold, environmental ethics, and the land ethic. *Wildlife Society Bulletin* 26: 742–744. Commemorative issue celebrating the 50th anniversary of "A Sand County Almanac" and the legacy of Aldo Leopold.
Norgaard, R. B. (1994). *Development Betrayed: The End of Progress and a Co-Evolutionary Revisioning of the Future*. Routledge, London.
Norton, B. G. (1984). Environmental ethics and weak anthropocentrism. *Environmental Ethics* 6(2): 131–148. DOI: 10.5840/enviroethics19846233 (Accessed April 6, 2022).
Norton, B. G. (2005). *Sustainability: A Philosophy of Adaptive Ecosystem Management*. University of Chicago Press, Chicago.
Perry, W. G. Jr. (1968). *Forms of Ethical and Intellectual Development in the College Years: A Scheme*. Rinehart and Winston Inc., Holt, New York.
Piccolo, J. (2010). Intrinsic values in nature: The objective good or simply half of an unhelpful dichotomy. *Journal for Nature Conservation*. DOI: 10.1.106j.j.in.c.2017.02 .007/ (Accessed April 22, 2022).
Pinker, S. (2008). The moral instinct. *New York Times Magazine* (Accessed January 2008). Available at: https://www.researchgate.net/publication/265066550_The_Moral_Instinct (Accessed March 17, 2022).
Price, T. D. and Yosef, O. B. (Eds.). (2011). The origins of agriculture: New data, new ideas. *Current Anthropology: The Wenner-Green Symposium Series* 52: 1–35.
Price-Williams, D. and Gaines, R. (1994). The dreamtime and dreams of Northern Australian aboriginal artists. *Ethos* 22(3): 373–388. DOI: 10.1525/eth.1994.22.3.02a00050 (Accessed March 15, 2022).
Rachels, J. (2003). *The Elements of Moral Philosophy* (4th edn.). McGraw-Hill, New York.
Raffalle, R., Robison, W. and Selinger, E. (2010). *Sustainability Ethics: Five Questions*. Automatic Press.
Rolston, H. III (1975). Is there an ecological ethics? *Ethics* 85(2): 93–109.
Rolston, H. III (1983). Values gone wild. *Inquiry* 26(2): 181–207.

Routley, R. (1973). Reprinted in Zimmerman, M. E. (Ed.). *Is There a Need for a New, an Environmental Ethic?* Proceedings of the XV World Congress of Philosophy, No. 1. Varna, Bulgaria. *Environmental Philosophy: From Animal Rights to Radical Ecology 1993* (Vol. 1973). Prentice Hall, Englewood Cliffs, NJ, pp. 12–21, 205–210.

Rowe, J. S. (1990). Ecocentrism: The cord that harmonizes humans and nature. *The Trumpeter* 4: 106–107.

Salner, M. (1986). Adult cognitive and epistemological development in systems education. *Systems Research* 3(4): 225–232.

Schlanger (2019). Quartz environment reporter. Available at: https://qz.com/1711172/the-global-demand-for-palm-oil-is-driving-the-fires-in-indonesia/amp/ (Accessed March 5, 2022).

Schröter, M., van der Zanden, E. H., van Oudenhoven, A. P. E., Remme, R. P., Serna-Chavez, H. M., de Groot, R. S. and Opdam, P. (2014). Ecosystem services as a contested concept: A synthesis of critique and counter-arguments. *Conservation Letters* 7(6): 514–523. DOI: 10.1111/conl.12091 (Accessed April 16, 2022).

Sinnot-Armstrong, W. (2019). Consequentialism. In: Zalta, E. N. (Ed.), *Stanford Cyclopedia of Philosophy*. Available at: https://plato.stanford.edu/archives/sum 2019/entries/consequentialism/ (Accessed March 6, 2022).

Skrimshire, S. (Eds.). (2010). *Future Ethics: Climate Change and Apocalyptic Imagination*. Continuum International Publishing Group, London.

Smith, M. (2001). *An Ethics of Place*. State University of New York Press, New York.

Steffen, W., Broadgate, W., Deutsch, L., Gaffney, O. and Ludwig, C. (2015). The trajectory of the anthropocene: The great acceleration. *The Anthropocene Review* 1: 18–36. DOI: 10.1177205301961456 4785/.

Talbert, M. (2019). Moral responsibility. In: *Stanford Encyclopaedia of Philosophy*.

Thompson, P. B. (1995). *The Spirit of the Soil* (1st edn.). Routledge, London.

Thompson, P. B. (1998). *Agricultural Ethics: Research, Teaching, and Public Policy*. Iowa State University Press, Ames.

Thompson, P. B. (2014). Agricultural ethics. In: Thompson, P. B. and Kaplan, D. M. (Eds.), *Encyclopaedia of Food and Agricultural Ethics*. Springer Reference, New York.

Thompson, P. B. and Kaplan, D. M. (2014). Introduction. P. B Thompson and D. M Kaplan. In: *Encyclopaedia of Food and Agricultural Ethics* (Vol. 2014). Springer Reference, New York.

UCS (2020). How soil erosion threatens of food and farm future (and what to do about). Available at: https://www.ucsusa.org/resources/how-soil-erosion-threatens-food-and-farms. Union of Concerned Scientists.

Warren, M. A. (1997). *Moral Status: Obligations to Persons and Other Living Things*. Oxford University Press, Oxford.

Wright, R. (1995). *The Moral Animal; Why We Are the Way We Are: The New Science of Evolutionary Psychology*. Pantheon Books/Random House, New York.

Xu, Y. and Usher, B. (Eds.). (2006). *Groundwater Pollution in Africa*. Taylor & Francis, London. Available at: https://www.booktopia.com.au/groundwater-pollution-in-africa-yongxin- (Accessed March, 2, 2022).

Yankelovich, D. (1991). *Coming to Public Judgement: Making Democracy Work in a Complex Society*. Syracuse University Press, Syracuse, NY.

Zimdahl, R. (2006). *Agriculture's Ethical Horizon*. Elsevier Academic Press, New York.

Chapter 8

Migrant labour in agriculture: ethical issues and challenges

Francesca Giarè, Patrizia Borsotto and Gabriella Ricciardi, Council for Agricultural Research and Economics (CREA), Italy

1 Introduction

The agri-food sector is characterized by significant market concentration and asymmetric power relations which can result in unfair business practices. This particularly affects farm workers who often lack bargaining power in dealing with employers. About 300–500 million paid workers are engaged in agriculture and food production worldwide, with women accounting for 70% of the workforce (International Labour Organization (ILO) and UN Food and Agriculture Organization (FAO)). A significant percentage of workers are migrants, engaged mainly in casual, seasonal and temporary agricultural work. These workers play a vital role in global food production but are often unable to secure a sufficient income and adequate nutrition for themselves and their families.

Since they are not citizens of the countries in which they work, and lack the same legal status, migrant workers are particularly vulnerable to exploitation. In addition, the wide range of nationalities and the often temporary, seasonal nature of the work they do makes it harder for migrant workers to organize collectively, making their negotiating power very weak (Moraes et al., 2012). Adverse effects on migrant workers include lack of or one-sided employment contracts, precarious employment, excessive working hours, undeclared employment, underpaid or unpaid work and insufficient access to health services and social security benefits. Migrant workers face numerous obstacles to securing basic labour and human rights, including the right to food, a living wage, decent employment conditions and formal labour contracts that

http://dx.doi.org/10.19103/AS.2023.0125.17

meet employers' legal obligations under national law or that of bodies such as the EU, as well documented by the UN publication: Right to Freedom of Association of Migrants and their Defenders (A/HRC/44/42, 2020) and Nelson et al. (2020). In addition, they often experience poor housing conditions and insufficient access to healthcare and social protection. This situation has been worsened by the Coronavirus pandemic (COVID-19) which has severely restricted movement across borders, migration and global mobility, creating a form of 'sanitary apartheid' (Augenstein and Macchi, 2021). Migrant women are particularly vulnerable to exploitation in low-paid, menial occupations in sectors such as agriculture and to physical and sexual abuse (Cvajner, 2018).

As a sector, agriculture suffers from high numbers of fatal work accidents, with the sector accounting for over 50% of total accidents at work across a range of industries (Merisalu et al., 2019; Palumbo and Corrado, 2020). Migrant farmworkers are most at risk of death or injury from working on farms (Schenker, 2010; Arcury et al., 2010; Arcury et al., 2020). They also face unsafe working conditions in food processing and packing facilities. Problems such as unfair terms of employment and unsafe working conditions facing migrant workers have been documented in EU countries which are meant to have robust labour laws protecting workers (Callejón-Ferre et al., 2015; Kilkey and Urzi, 2017; Rye and Scott, 2018; Sargeant and Tucker, 2009). However, there is a need for more empirical research on the actual conditions of employment, health and safety protection experienced by migrant farmworkers as well as their own attitudes to their well-being, health and safety.

There are now established policies and systems for responsible business practices across global food supply chains. These are driven in part by increasing demands by consumers for ethically produced food products, including fair conditions for workers involved in growing and processing food products. However, these practices are still not mainstream within agri-food businesses globally. In addition, robust systems for verification and due diligence are not always strong, making it hard to assess the degree and effectiveness of implementation in practice. A broader issue is the social justice theory of ethics which highlights the multi-faceted nature of reality and therefore focuses on counterbalancing existing power structures by focusing specifically on the views of marginalized groups such as migrants (Mertens, 2015). A key question is how far the design and implementation of policies to support migrants should take on board their views and perceptions of their needs.

2 Importance of migrant workers in the agricultural and agri-food sectors

There were an estimated 173 million people living outside their country of birth or citizenship in 2000, reaching 221 million in 2010 and 281 million in

2020, an increase of over 60% in 20 years (UNDESA). In general, the number of international migrants grew faster than the global population: the share of international migrants in the total population rose from 2.8% in 2000 to 3.6% in 2020. In the same 20 years, the number of persons displaced across international borders while fleeing conflict, persecution, violence or human rights violations doubled from 17 to 34 million.

According to UNDESA, about two-thirds of international migrants live in just 20 countries, most notably the United States of America, home to about 51 million migrants, Germany, home to about 16 million people from other countries, and North Africa and West Asia, with a total of almost 50 million migrants. Europe has 87 million migrants, although many of these are European citizens moving within the EU Schengen area (i.e. without normal border controls between member states). India is the main source of migrants with about 18 million citizens who have decided to move outside their country of origin. Other countries with a large transnational community are Mexico and the Russian Federation (11 million each), China (10 million) and Syria (8 million). Most international migrants come from middle-income countries, while only 13% come from low-income countries, although the share of these has increased significantly over the past 20 years as humanitarian crises have grown in many areas of the world (Fondazione Migrantes, 2021). In 2020, almost half of all international migrants resided in their region of origin, with Europe leading in intra-regional migration: 70% of European-born migrants reside in another European country.

Migrant workers make up 13% of the EU's overall workforce. Certain sectors of the European economy, in particular the agri-food (including agriculture and food and drink processing) and tourism sectors, depend on the support of seasonal workers from EU and non-EU countries for specific periods of the year, equal to about 1 million people per year (European Commission, 2021). In 2020, non-EU nationals employed as 'agriculture and fisheries' workers represented 2.6% of the EU labour force with only 0.6% EU nationals. In the care sector, non-EU nationals are also over-represented: 5.6% of non-EU workers are employed as personal care workers; this is only the case for 2.9% of EU workers.

Estimates indicate that the dependency of EU countries on migrants to perform agricultural and care work is likely to increase in the future (Hogarth, 2021). Factors such as relatively poor pay, unsocial hours and challenging working conditions (which make such work unattractive to many), an ageing European population, and demand for agricultural products and personal services suggest the demand for labour in the agricultural and care sector will only increase. If the current trend of low local labour force participation in these sectors remains constant, more migrants will be needed to fill labour gaps. Future labour needs in the agriculture and care sectors will therefore require policymakers to develop informed, effective and socially cohesive integration policies for the increasing number of migrants that will be employed in these sectors.

Italy constitutes a particularly interesting case among European countries. Geographically, Italy lies at the centre of the Mediterranean, which has meant that it has always been at the centre of the movement of peoples: significant internal migration and emigration have characterized the history of the country. Towards the end of the 1970s, the country experienced a relatively stable pattern of immigration, in part as a transit area for the movement of migrants to the countries of Northern Europe. According to the latest available data (ISTAT, 2021), there are about 5 million foreigners living in Italy, equal to 8.5% of the resident population. Looking at those in employment, Fig. 1 shows a steady increase in the percentage of foreign workers as a percentage of the total workforce in Italy to just under 11%, a rise that was interrupted by the COVID-19 pandemic that began in 2019. As with other countries, labour shortages as a result of the COVID-19 emergency have highlighted the dependence of the Italian agri-food sector on cheap and flexible migrant labour (Open Society Foundation, 2020).

A recent study by the National Rural Network (RRN, 2020) found that 40% of foreign residents live in urban centres (2.1 million) and 60% (3.1 million) live in rural areas which have seen a significant increase (around 73%) in the last decade. In rural areas, often with fewer services and poorer transport infrastructure, the cost of living is lower. While Italians have tended to abandon these areas for urban centres, migrants tend to settle in cheaper rural locations where they work in sectors such as agriculture.

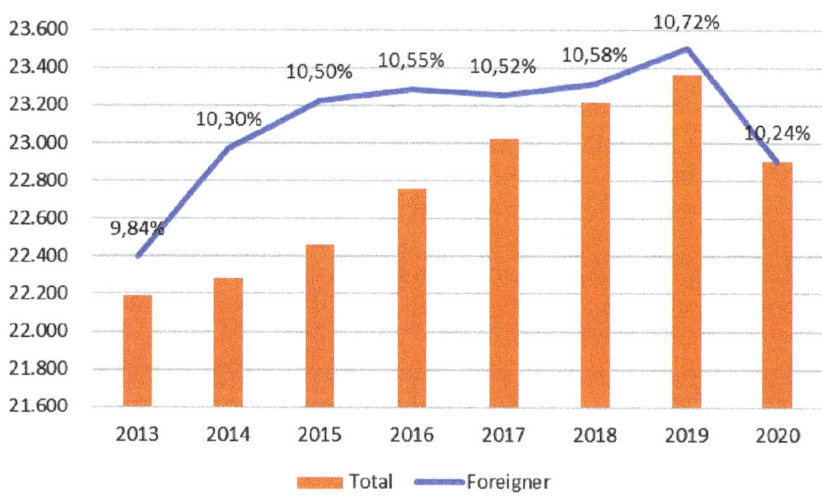

Figure 1 Total number of employed persons (.000) and foreign employment rate in Italy. Years 2013–2020. Source: Adapted from ISTAT.

3 Migrants' rights in Europe

Migration to and within Europe is governed by a complex set of rules, which includes national law, EU law, the European Convention on Human Rights (ECHR), the European Social Charter (ESC) and a series of other international obligations assumed by member states.

EU law classifies third-country nationals into 25 different categories, including persons in need of protection (asylum seekers, beneficiaries of subsidiary or temporary protection, refugees and victims of trafficking) and illegal migrants (third-country nationals whose stay is irregular and whose planned removal, in some cases, has been postponed). Each of the 25 categories guarantees different rights, which vary according to the links they have with the member states, or which arise from their need for special protection (FRA, 2020). EU law governs only certain categories of third-country nationals, such as asylum seekers, in detail, leaving member states a wide margin of discretion in how they deal with other types of migrants (FRA, 2020).

Two common and interlinked features of many national policies relating to migrants are:

- whether migrants enter a country as part of a programme planned and managed by the state or do so outside government regulation and control; and
- the perceived usefulness of migrants e.g. in filling gaps in the labour market.

These issues of legality and perceived value of a migrant's presence in a country can have a profound effect on how a migrant is treated. A key ethical challenge for both states and individual citizens is how fair these criteria are in the context of the broader framework of human rights.

For most migrants, access to employment, education, housing, health care, social security, social assistance and other social benefits can be particularly complex and generally depends on the recognition of the 'Right to stay'[1] (FRA, 2020). Regarding employment, the principle of equal treatment, which prohibits any discrimination based on racial or ethnic origin, has been enshrined in EU law.[2] The fundamental social rights of workers were enshrined in the Community Charter of 9 December 1989 and further developed in the EU Charter of Fundamental Rights, even if they are few and essentially concern citizens and/or legally resident persons in a country. As a rule, access to the

[1] For third-country nationals, a single application procedure for the issue of a 'single permit' was introduced in 2011, which allows them to reside and work in the territory of a member state and also provides for a common set of rights in the case of workers who are legally resident in a member state (Single Permit Directive (2011/98/UE)).
[2] This principle, enshrined in the Equal Treatment Directive (2000/43/EC), also covers access to goods and services, social protection and the social security system.

labour market is strictly dependent on the legal status of a person. For this reason, in September 2020 the European Commission (EC) adopted a package of regulatory measures: the 'New Pact on Migration and Asylum', to develop, inter alia, new procedures aimed at rapidly establishing status on arrival and a new common framework for solidarity and shared responsibility in dealing with migrants.

Predating the foundation of the EU is the Council of Europe which was set up in the aftermath of the Second World War to uphold human rights and the rule of law in Europe. EU member states are members of the Council which means they must abide by the terms of the ECHR and the jurisdiction of the ECHR. Within the Council of Europe, basic human economic and social rights are governed by the ESC. This complements the ECHR by establishing a monitoring mechanism to ensure compliance with its provisions. The provisions of the ESC are also extended to migrants in an irregular situation which is monitored by the European Committee of Social Rights.

This framework means that all workers have fundamental rights, including the right to be paid for work, regardless of whether they are legally employed or not. The economic and social rights expressly guaranteed by the ECHR concern the prohibition of slavery and forced labour (Article 4), as well as the right to participate in the establishment of trade unions (Article 11). Other protection provisions concern legal migrant workers and ensure, for example, that loss of employment doesn't automatically result in the loss of a residence permit, giving the person the opportunity to look for another job. The ECHR expressly recognizes several rights for migrants such as the right to protection from unjustified dismissal, fair and equitable working conditions and the right to health and safety at work (FRA, 2020).

In order to give greater attention to respect for human rights within the business world, on 23 February 2022 the EC published a proposal for a Directive on Corporate Sustainability Due Diligence, in response to the resolution of March 2021 by the European Parliament to establish a common framework for due diligence obligations by companies. The proposed directive intends to create a common obligation of due diligence by employers which includes human rights. According to the proposal, companies will have to implement systems and processes aimed at identifying, preventing and mitigating possible infringements of human rights in their supply chains. The companies affected by the proposed directive include those operating in sectors such as agriculture. The proposed directive defines human rights violations in terms of the international conventions listed in the relevant annex, including those on child labour, exploitation of workers, discrimination etc. These obligations would extend to company contracts and relationships with their suppliers since suppliers and sub-contractors are more likely to use migrant labour. Sanctions are envisaged for companies that do not take corrective measures against failures of due diligence, including excluding companies from public tenders (Augenstein and Macchi, 2021).

This complex framework of legislation at European level becomes even more complex when it intersects with national legislation and government policy, as the example of Italy illustrates (Borsotto et al., 2022). Italian public policies on immigration were established in 1998 by Law 40/1998, known as the 'Turco-Napolitano Law', which fixed annual entry quotas for various categories of foreign workers. In 2011, this resulted in greatly reduced quotas for non-Italian workers which were almost halved for seasonal ones (Corrado et al., 2018).

This legislation was complemented by the 'Consolidated Act on Immigration and the Status of Foreigners' (Legislative Decree 25 July 1998, no. 286), which has maintained its general structure despite numerous changes. It governs three main areas of immigration:

1 The right of immigration;
2 The general management of migration (e.g. rules governing entry and the right to stay); and
3 The right of integration i.e. rules governing extension to migrants of civil, social and political rights enjoyed by Italian citizens.

The fundamental principles underlying the Consolidated Act are control of migratory flows, reducing illegal immigration and the granting of a wide range of rights aimed at integration of foreigners with a legal right to stay. The Italian legal system is characterized by the commitment to promote the reception and integration of immigrants. Immigrants are guaranteed the fundamental rights of freedom and equality established by the first part of the Italian Constitution, including the right of asylum expressly provided for in favour of foreigners by Article 10. Full citizenship can be obtained e.g. through birth, marriage or naturalization.

The most significant recent regulatory interventions have been Legislative Decree 215/2003 and Legislative Decree 216/2003.[3] These include general provisions to prevent discrimination in the workplace. In compliance with the provisions of European Law 2019/2020,[4] the 'nationality' factor has been added to the prohibited areas of discrimination listed in Legislative Decree 216/03 (age, sexual orientation, religion, personal beliefs, disability). The

[3] The D. Lgs. 216/2003, 'Implementation of Directive 2000/78/EC on equal treatment in employment and occupation (and of Directive 2014/54/EU on measures to facilitate the exercise of the rights conferred on workers in the context of the free movement of workers. Implementation of workers)' to Article 1 reads: This Decree contains provisions on the implementation of equal treatment for persons irrespective of religion, belief, disability, age (nationality) and sexual orientation, with regard to employment and working conditions, providing for the necessary measures so that these factors do not cause discrimination, in a way that also takes into account the different effect that the same forms of discrimination can have on women and men.

[4] Law No. 238 of 23 December 2021, containing 'Provisions for the fulfilment of obligations arising from Italy's membership of the European Union - European Law 2019-2020'.

legislative decree has been extended beyond the field of work as required by EU Directive 2000/78/EC.

In practice, these policies in controlling illegal immigration have proved hard to enforce and have failed to meet labour requirements in sectors such as agriculture. This has resulted in the use of migrants from Eastern European member states (with greater freedom of movement within the Schengen area), undocumented workers and, since 2015, asylum seekers and non-EU refugees (Corrado et al., 2018). These less regulated and sometimes illegal flows of workers make abuse of migrant workers in sectors such as agriculture all too easy for unscrupulous employers.

These problems led in 2020 to the approval of a 'National Action Plan to Tackle Labour Exploitation, Unlawful Recruitment and Forced Labour in Agriculture (2020–2022)'. Agreed by various ministries (including Labour, Agriculture and Food, Justice and the Interior), the plan seeks to deal with problems such as illegal recruitment and undeclared work, occupational health and safety, and poor living conditions. It focuses on the four pillars of prevention, protection, enforcement and remedies for poor practices. Following on from this, 'National guidelines on identification, protection and assistance to victims of labour exploitation in agriculture' were approved in October 2021. They highlight the importance of creating a regional network of protective measures based on a mix of public and private bodies, including professional groups such as social workers. Italy's commitment to ensure greater respect for the rights of agricultural workers is also reflected in the proposal for a law on 'ethical' agri-food chains (Bill, 1373/2021). This offers consumers the means to make informed food choices and provides for tax incentives to reward ethical agri-food businesses. Legislative Decree 91/2014 also supports businesses that respect fair working and employment conditions in agriculture (Corrado and Palumbo, 2022). Effective integration of migrants requires this kind of broad-ranging and systematic approach (Hampshire, 2013).

4 European agricultural policies and migrant labour

According to the Treaty on European Union (EU) of 1993 (the Maastricht Treaty), the EU 'shall promote economic, social and territorial cohesion, and solidarity among Member States', as well as 'shall combat social exclusion and discrimination'. The Treaty, in addition, states that it 'shall work for the sustainable development of Europe based on balanced economic growth and price stability, a highly competitive social market economy, aiming at full employment and social progress, and a high level of protection and improvement of the quality of the environment' (Consolidated version of the Treaty on European Union). The EU Multiannual Financial Framework is an investment budget that 'pools resources to implement policies and brings an added value to the benefit of all EU citizens by delivering on common challenges such as fighting climate change and environment protection, digital challenges, defense and border security, social rights and jobs' (European Parliament, 2020).

The Common Agricultural Policy (CAP) is one of the oldest EU policies and accounts for almost 40% of the entire EU budget. Post-2020 CAP must face new challenges to meet a range of sustainability goals. There are nine objectives, three of which relate to the environment and climate. CAP is implemented through national strategic plans, including policy targets, and allocation of payments. The main change introduced by the CAP post-2020 is the 'new delivery model' for payments, which connects financing with achievement of environmental and social objectives (Núñez Ferrer et al., 2018; Chereji et al., 2020). For the first time, the CAP includes social conditionality, meaning that CAP beneficiaries must respect European social and labour law to receive CAP funds.[5] Ethical sustainability is a recent theme that broadens the scope of the concept of 'sustainability' to include issues of equity and fairness in rural communities (Schmitt et al., 2016). Those who do not comply with these social requirements risk losing funding.

A series of Agricultural Knowledge and Innovation System (AKIS) hubs (bringing together farmers, government officials, experts, researchers and others) will ensure farmers have access to information on how to comply with the social dimension of the CAP. A social dimension aimed at improving the conditions for agricultural workers will also be added to programmes supporting particular sectors such as fruit and vegetable, olive and hop cultivation. Dedicated EU resources and specific interventions (i.e. AMIF[6]) are also available specially to help protect migrant workers. The EC is mandated to evaluate the effectiveness of this social dimension by 31 December 2026 and propose ways it can be improved. It has highlighted the need to strengthen national measures to prevent illegal hiring and labour exploitation. A potential issue with CAP is that it is focussed on correcting distortions in agricultural markets (e.g. moving away from a single-minded focus on production) and improving overall performance at the farm level. A challenge will be to ensure that the rights of farmworkers, and migrant workers in agriculture, receive sufficient attention.

[5] The agreement on the financing, management and monitoring of the Common Agricultural Policy (2 December 2021) stipulates that financial support from the CAP will have to comply with major EU regulations regarding workers, in particular, Directive 89/391/EEC of the Council of 12 June 1989 on the introduction of measures to encourage improvements in the safety and health of workers; Directive 2009/104/EC of the European Parliament and of the Council of 16 September 2009 on the minimum safety and health requirements for the use of work equipment; Directive (EU) 2019/1152 of the European Parliament and of the Council on transparent and predictable working conditions in the European Union. In addition, the Commission will have to assess the feasibility of adding to this list, starting from 1 January 2025, regulation 492/2011 on the free movement of workers within the Union.

[6] The Asylum, Migration and Integration Fund (AMIF) is set up for the period 2021–2027, with a total of EUR 9.9 billion. The Fund aims to further boost national capacities and improve procedures for migration management, as well as to enhance solidarity and responsibility sharing between member states, in particular, through emergency assistance and the relocation mechanism (https://home-affairs.ec.europa.eu/funding/asylum-migration-and-integration-funds/asylum-migration-and-integration-fund-2021-2027_en).

5 Conclusion

Current approaches to supporting migrant workers within the EU are still piecemeal and, most recently, have been prompted by immediate labour shortages in the aftermath of the COVID-19 pandemic. There is still a tendency to treat migrant workers purely as an economic resource to fill temporary gaps in the labour market in sectors such as agriculture. This both ignores migrant workers' own views on what they need and the challenges they face as well as preventing a more holistic and sustainable approach to meeting those needs.

Given the opportunity, most non-EU citizens prefer to secure a residence permit in advance of taking on work which then gives them a legal status and consequent access to basic rights enjoyed by full citizens of a country. A starting point for better treatment is fair and accessible rules and systems to allow immigrants to acquire a valid residence permit in order to achieve a degree of equality between citizens and foreigners. However, this needs to be backed up by broader, properly integrated interventions aimed at guaranteeing safe and legal entry systems for workers with low and medium qualifications, strengthening wages and workers' rights, guaranteeing decent living conditions, developing welfare services and supporting fair and sustainable agri-food chains among those employing migrant workers or using sub-contractors who do (Corrado and Palumbo, 2022). These programmes will only succeed if there is a fuller picture of migrant labour and the conditions in which it operates in sectors such as agriculture.

6 References

Arcury, T. A., Estrada, J. M. and Quandt, S. A. (2010). Overcoming language and literacy barriers in safety and health training of agricultural workers. *J. Agromed.* 15(3), 236–248. https://doi.org/10.1080/1059924X.2010.486958.

Arcury, T. A., Quandt, S. A., Arnold, T. J., Chen, H., Sandberg, J. C., Kearney, G. D. and Daniel, S. S. (2020). Work safety culture of Latinx child farmworkers in North Carolina. *Am. J. Ind. Med.* 63(10), 917–927. https://doi.org/10.1002/ajim.23161.

Augenstein, D. and Macchi, C. (2021). *The Role of Human Rights and Environmental Due Diligence Legislation in Protecting Women Migrant Workers in Global Food Supply Chains* (May 17, 2021). Research Policy Study commissioned by Oxfam Germany and Action Aid France in the framework of the EU DEAR Project 'Our Food. Our Future', 2021, Available at SSRN: https://ssrn.com/abstract=3847676 or http://dx.doi.org/10.2139/ssrn.3847676.

Borsotto, P., Giarè, F. and Ricciardi, G. (2022). Immigrazione: le opportunità offerte dall'agricoltura sociale. In: Borgi, M., Genova, A., Collacchi, B. and Cirulli, F. (A cura di). *Rapporti ISTISAN 22/9. Agricoltura Sociale: Processi, Pratiche e Riflessioni per l'Innovazione Sociosanitaria, Rome, Istituto Superiore di Sanità, 2022. (Rapporti ISTISAN 22/9).* ISSN: 1123-3117 (Cartaceo) - 2384-8936 (online).

Callejón-Ferre, A. J., Montoya-García, M. E., Pérez-Alonso, J., and Rojas-Sola, J. I. (2015). The psychosocial risks of farm workers in south-east Spain. *Saf. Sci.* 78, 77–90. https://doi.org/10.1016/j.ssci.2015.04.015.

Chereji, A. I., Chiurciu, I. A., Maerescu, C. M., Țuțui, D. and Chereji, I. Jr. (2020). Political and legal conditionalities for the cohesion policy and CAP post-2020. Between real issues and unjustified fears. In: Palumbo, L. and Corrado, A. (Eds), *COVID-19, agri-Food Systems, and Migrant Labour: the Situation in Germany, Italy, The Netherlands, Spain, and Sweden*. Oradea, Open Society European Policy Institute (OSEPI).

Corrado, A., Caruso, F. S., Lo Cascio, M., Nori, M., Palumbo, L. and Triandafyllidou, A. (2018). *Is Italian Agriculture a 'Pull Factor' for Irregular Migration – and, If so, Why?* Oradea, Open Society European Policy Institute (OSEPI).

Corrado, A. and Palumbo, L. (2022). Essential farmworkers and the pandemic crisis: migrant labour conditions, and legal and political responses in Italy and Spain. In: Triandafyllidou, A. (Ed.), *Migration and Pandemics*. Cham: Springer, 145–166.

Cvajner, M. (2018). *Sociologia delle Migrazioni Femminili. L'Esperienza delle Donne Post-Sovietiche, Bologna, Il Mulino, Brossura*. ISSN: 0486–0349.

European Commission (2021). Statistics on migration to Europe. European Commission (European Union). Available at: https://ec.europa.eu/info/strategy/priorities-2019-2024/promoting-our-european-way-life/statistics-migrationeurope_en#employment-of-immigrants (accessed 29 September 2021).

European Commission (2022). Proposal for a Directive of the European Parliament and of the Council on Corporate Sustainability Due Diligence and amending Directive (EU) 2019/1937. COM(2022) 71 final.

European Parliament (2020). Q&A on the EU's long-term budget (Multiannual Financial Framework, MFF). Available at: https://www.europarl.europa.eu/news/en/press-room/20200115ipr70326/q-a-on-the-eu-s-long-term-budget-multiannual-financial-framework-mff (accessed 10 November 2020).

European Union Agency for Fundamental Rights and Council of Europe (2020). *Handbook on European Law Relating to Asylum, Borders and Immigration*. Luxembourg: Publications Office of the European Union, 2020. ISBN 978-92-871-9800-6; FRA – print: ISBN 978-92-9474-985-7; FRA – web: ISBN 978-92-9474-977-2.

Fondazione Migrantes (2021). XXX rapporto immigrazione 2021. Verso Noi Sempre Più Grande.

Hampshire, J. (2013). *The Politics of Immigration*. Cambridge: Polity Press.

Hogarth, T. (2021). Covid-19 and the demand for labour and skills in Europe early evidence and implications for migration Polic. Migration Policy Institute. Available at: https://www.migrationpolicy.org/sites/default/files/publications/mpie-hogarth_covid19-labour-final.pdf.

Kilkey, M. and Urzi, D. (2017). Social reproduction in Sicily's agricultural sector: Migration status and context of reception. *J. Ethn. Migrat. Stud.* 43(15), 2573–2590. https://doi.org/10.1080/1369183X.2017.1286971.

Merisalu, E., Leppälä, J., Jakob, M. and Rautiainen, R. H. (2019). Variation in Eurostat and national statistics of accidents in agriculture. *Agron. Res.* 17(5), 1969–1983. https://doi.org/10.15159/AR.19.190.

Mertens, D. M. (2015). Philosophical assumptions and program evaluation. *Spazio Filos.* 13, 75–85.

Moraes, N., Gadea, E., Pedreño, A. and De Castro, C. (2012). Enclaves globales agrícolas y igraciones de trabajo: convergencias globales y regulaciones transnacionales. *Política y Sociedad* 49, 13–34.

Nelson, V., Martin-Ortega, O. and Flint, M. (2020). Making human rights due diligence work for small farmers and workers in global supply chains. *Fairtrade Advocacy Office and Bort fur Deiwelt, 2020. Available online:* https://fairtrade-advocacy.org/wp-content/uploads/2020/06/UoG-HRDD-Full-Report-60pp-FINAL-SECURED.pdf.

Núñez Ferrer, J., Musmeci, R., Rinaldi, D. and Polli, O. (2020). *The EU Budget and Its Conditionalities: an Assessment of Their Contribution to Performance and Reforms*. Brussels: CEPS. Available at: https://www.ceps.eu/ceps-publications/the-eu-budget-and-its-conditionalities/ (accessed 3 November 2020).

Open Society Foundation (2020). Covid-19, Agri-food systems, and migrant labour. The situation in Germany, Italy, The Netherlands, Spain, and Sweden. Open Society Foundation. Available at: https://www.opensocietyfoundations.org/uploads/ccf241cc-89b2-4b44-a595-90bd77edab3d/covid19-agrifood-systems-and-migrant-labour-20200715.pdf.

Palumbo, L. and Corrado, A. (2020). *COVID-19, agri-Food Systems, and Migrant Labour: the Situation in Germany, Italy, The Netherlands, Spain, and Sweden*. Open Society European Policy Institute (OSEPI).

Rye, J. F. and Scott, S. (2018). International labour migration and food production in rural Europe: A review of the evidence. *Sociol. Ruralis* 58(4), 928–952. https://doi.org/10.1111/soru.12208.

Sargeant, M. and Tucker, E. (2009). Layers of vulnerability in occupational safety and health for migrant workers: case studies from Canada and the UK. *Policy Pract. Health Saf.* 7(2), 51–73. https://doi.org/10.1080/14774003.2009.11667734.

Schenker, M. B. (2010). A global perspective of migration and Occupational Health. *Am. J. Ind. Med.* 53(4), 329–337. https://doi.org/10.1002/ajim.20834.

Schmitt, E., Keech, D., Maye, D., Barjolle, D. and Kirwan, J. (2016). Comparing the sustainability of local and global food chains: a case study of cheese products in Switzerland and the UK. *Sustainability* 8(5), 419 https://doi.org/10.3390/su8050419.

UNDESA (2021). *International Migration 2020 Highlights*. Available at: www.unmigration.org.

Zumpano C. (Ed.) (2020). *Migrazioni, Agricoltura e Ruralità. Politiche e Percorsi per lo Sviluppo dei Territori, Rapporto di Ricerca*. Rome, CREA-RRN.

Chapter 9

Sustainable intensification in agriculture: ethical issues

Alesandros Glaros, Philip Quarshie, Emily Duncan and Evan Fraser, University of Guelph, Canada

1 Introduction

Sustainable intensification (SI) is heralded as a key driver for food systems' transformation by prominent organizations, including the Food and Agriculture Organization of the United Nations (FAO, 2016), the World Bank Group (World Bank Group, 2022), and the International Maize and Wheat Improvement Center (Donovan, 2020). SI is an 'umbrella term' (Pretty and Bharucha, 2014, p. 1580), referring to an array of tools, techniques, and approaches to increase production on existing agricultural land while enhancing or, at minimum, without compromising ecosystem services (Rockström et al., 2017). SI is born from an argument that current strategies for producing food sustainably are 'not enough' (Tilman et al., 2011). Business-as-usual scenarios for food production surpass emissions reduction targets and other 'planetary boundaries' and may fail to adequately contribute to healthy diets and address food insecurity globally (Rockström et al., 2017). An array of social, political, economic, and biophysical factors indicates a need for sustainably producing more food, globally, on existing farmland (Cassman and Grassini, 2020). SI places a particular focus on increasing yields of smallholders who produce an estimated 30–34% of the global supply on land under two hectares (Ricciardi et al., 2018). Indeed, a priority for global agricultural development initiatives is addressing smallholder yield gaps (e.g. Van Ittersum et al., 2016). Proponents of SI argue we must achieve 'win-win-wins' of social, environmental, and

http://dx.doi.org/10.19103/AS.2023.0125.09

economic sustainability when it comes to agri-food production, contributing to smallholder incomes and livelihoods, all while minimizing environmental effects of agriculture. Key dimensions of SI research include linking the cross-scale effects of individual farm SI practices on the regional food system and ecosystem outcomes; addressing yield gaps in sub-Saharan Africa and Asia; cropping and livestock system diversification; and establishing metrics to measure SI (Cassman and Grassini, 2020).

There has been a substantial debate on the definition of SI and policy/programme responses to pursue it. Many are critical of the term, perceiving a lack of critical attention to measures for social or economic sustainability (Mahon et al., 2018). Fiercer critiques are from proponents for agroecology, who argue the term 'greenwashes' approaches to food system transformation, reinforcing top-down, technocratic, corporate-dominated models of production. Debates go back many decades. The 2009 'IAASTD' report suggested that technologically focused solutions of the world food crisis may simply be a Trojan horse that undermines small-scale farmers at the expense of benefits for large corporations. Today, these debates remain unchanged although the potential for technology to promote sustainable development has arguably increased. For example, the so-called digital agricultural revolution offers the promise of using advanced technologies such as artificial intelligence and robotics, and other technologies, to produce more food on less land and with a smaller environmental footprint (Green et al., 2021). Nevertheless, and as will be discussed below, these claims remain contested. Ensuing debates over SI centre on several dimensions, including how to feed the world sustainably, and what ability/opportunity exists to scale production to meet global food demand (Bernard and Lux, 2017). The term SI has evolved in response to its critiques, incorporating the pursuit of social and economic capital as key goals, thereby developing into a more 'holistic' measure for agroecosystem performance (Mouratiadou et al., 2021; Smith et al., 2017)

A suite of emerging food production technologies: cellular agriculture, digital agriculture, and controlled environmental agriculture (= 'agriculture 4.0 technologies') promise to fulfil the goals of SI in novel and radical ways (see, e.g. Glaros et al., 2021; Fraser and Campbell, 2019). Climate-smart agriculture can be perceived as a corollary to SI (Campbell et al., 2014). There is, on one hand, substantial optimism regarding these new methods for SI. Yet there is scepticism and cautionary arguments that these new tools serve to make the Trojan horse even 'bigger' and 'uglier' than before. The pace of these technologies' evolution, development, and commercialization makes revisiting the debate over SI timely and prescient.

In this chapter, we present the 'imperative' for SI and discuss why it is invoked as a necessary food security approach. Following this, we adopt the premises and policies framework presented by Garnett (2014) to sketch and review the ethical dilemmas facing SI, highlighting the contribution of agriculture 4.0 technologies to these debates (Fig. 1). Our analysis examines each of these policy imperatives in turn and highlights the key ethical trade-offs that comprise each. We highlight SI's roots in environmental pragmatism and recommend 'pragmatic responses' to negotiate the tensions and trade-offs outlined in the previous section, highlighting relevant case studies. This analysis contributes to bridging the dichotomous thinking that has divided food security debates focused on SI and offers key principles to negotiate the tensions that are prevalent in food system transformations.

2 The imperative for sustainable intensification

SI is proposed by some sections of the scientific community to increase crop and livestock yields and reduce agriculture's footprint on the environment, with attention to soil and water resources and the integrity of ecosystems (Garnett et al., 2013; Godfray, 2015; Tilman et al., 2011; Godfray and Garnett, 2014; Cassman and Grassini, 2020). The underlying premises on which SI is proposed are as follows:

1. The need for increased production;
2. Increased production must be met through higher yields rather than bringing new land into production;
3. Managing sustainability; and
4. Using a range of diverse approaches (Garnett et al., 2013).

In what follows, we discuss these underlying SI premises in further detail. Overall, SI is an approach to boost agricultural yield without converting new land to production, contributing to farmer livelihoods while reducing negative externalities to the environment and animal welfare (Pretty and Bharucha, 2014).

The focus on the SI approach is driven by the challenge of how to feed a growing global population. Demand for food will rise over the coming decades as the world population increases towards 9.8 billion in 2050. The 2013 World Development Report by the World Bank projects that by 2030 worldwide demand for food will increase by 50%, and for meat by 85% (Evans, 2011). Food supply challenges are enmeshed within complex systems involving energy insecurities, fuel and fertilizer production, transportation costs, biofuels, geopolitics, climate, and a host of other systems (Homer-Dixon et al., 2015). Water scarcity as a result of prolonged drought - an externality of climate change and unsustainable extraction of water from rivers, lakes, and groundwater - is also affecting food production. The 2018 United Nations (UN) World Water Development Report (WWDR) predicts that nearly 6 billion people will suffer from clean water scarcity by 2050 (WWDR, 2018), a situation that will affect food production as agriculture accounts for 70% of global freshwater use, forcing many nations, particularly those in developing regions such as sub-Saharan Africa to be net importers of food (Evans, 2011). To avert possible global food supply challenges in the near future, some scholars argue that world food production must increase dramatically. The food production increment needed to address food shortage by 2050, some argue, is between the ranges of 70% and 100% of what is currently produced (Pretty et al., 2010; Conway, 2012; Hunter et al., 2017).

The first and second bedrock underlying SI assumptions is that increased production must be met through higher yields because increasing the area of land in agriculture carries major environmental costs. With agricultural lands now being lost to urbanization, desertification, salinization, and soil erosion, the conversion of remaining lands (which consist mainly of forests, wetlands, or grasslands) would greatly increase greenhouse gas (GHG) emissions and the loss of biodiversity and important ecosystem services (Tscharntke et al., 2012; Tilman et al., 2002). Therefore, scholars argue that it is imperative to increase yields on existing and/or underperforming farmlands, particularly for low-income developing countries of sub-Saharan Africa and Asia where current yield gaps are large (Cassman and Grassini, 2020; Pretty et al., 2010). It is important to recognize that, despite these challenges and regional nuances, SI has generated marked results across global agricultural land-use metrics. The past several decades have seen a global-level plateau in agricultural land use, mainly driven by decreases in the amount of global pastureland (Taylor and Rising, 2021). All the while, yields have continued to increase, albeit with some signs of plateauing (Grassini et al., 2013). The rhetoric of increasing yields or closing yields gaps is often described (and indeed critiqued) as best achieved through high-input industrial agriculture models, ranging from high-yielding seeds, fertilizer, and agrochemicals, mechanization through to biotechnology or genetically modified organisms (GMOs) for food production (Godfray et al., 2010). In some regions, scholars argue that survival through drought, wars, and other major causes of famine can only be achieved by embracing these technologies to produce better food with less effort (Juma, 2011).

A key part of the discourse around global food security is the challenge of how reconciling the need to achieve these food production targets without environmental trade-offs because the agricultural sector is widely held responsible for cross-scale environmental changes (Rockström et al., 2017). The IPCC 2022 report suggests a more vigorous, robust, and urgent approach to reducing climate-changing emissions and reduction of average global temperature to well below +1.5°C pre-industrial level. This means agriculture, as a major contributor to greenhouse gas emissions, must play its part in addressing and solving the problem. Current estimates suggest that approximately 15% of all GHGs emissions come from food production (about the same amount comes from land conversion). Any serious attempt to reduce the rise in global temperature must include the agriculture sector, a move that gives SI a dual role of ensuring food security and environmental sustainability (Garnett et al., 2013; Godfray and Garnett, 2014).

The last underlying premise for SI is the approach must go beyond a single food production and distribution practice to diverse place-based, context-specific food production and distribution practices that ensure sustainability and environmental integrity (Garnett et al., 2013; Godfray and Garnett, 2014). According to Garnett et al. (2013), these constellations of diverse food production approaches may include but are not limited to the use of biotechnology, organic, digital, and agroecological practices and methods. This argument that agriculture and food systems are diverse and heterogeneous across scales, particularly smallholding agriculture in sub-Saharan Africa, led to the reclassification of SI into three sub-categories: Genetic Intensification, Ecological Intensification, and Market Intensification (Conway, 2012). Genetic intensification includes plant and animal breeding through genetic modification. Ecological intensification entails making far more effective use of resources and applying principles from the organic and agroecological movements to boost productivity. Market intensification is the process of creating an enabling economic environment that allows farmers to access investment money and markets for their output in order to encourage improved production (Conway, 2012; Godfray, 2015). The sub-categories of SI help to ensure that there are a diverse range of solutions that will assist in food system transformation.

Despite these underlying premises, the term SI is fiercely contested and has many criticisms from food security scholars regarding some of its approaches to addressing agriculture's effects on the environment. Some argue SI is a Western model of agriculture that, when introduced to developing nations, will harm smallholder agriculture (e.g. Altieri and Nicholls, 2012; Holt-Giménez et al., 2012; Tomlinson, 2013). Others openly describe SI as a method to sneak in intensive farming under the camouflage of sustainability (Collins and Chandrasekaran, 2012), with widespread and serious ethical issues relating to animal and plant breeding. This move has prompted the research and policy community to identify four policy areas which need to be given attention when implementing SI that are detailed in the next section.

3 Policy goals and ethical trade-offs

Having established the imperative for SI and outlined some of its controversies, we now draw out some of its policy goals as defined by Garnett (2014). These policy goals are broad areas of agri-food systems that clearly interface with SI. While these are described as goals, they also hold ethical trade-offs in the sense that achieving one gains in one aspect may lead to negative consequences for another. This paradox creates the need for agri-food researchers and policymakers to more deeply consider the ethical dilemmas embedded in SI. The goals broadly include biodiversity and land-use conservation, animal and livestock welfare, health and nutrition, and sustainable development and rural economies. We review SI approaches to achieve each of these goals and, in so doing, we define what we consider to be the key underlying ethical tensions. We conclude by summarizing these ethical trade-offs, reflecting on how technologies of agriculture 4.0 advance novel considerations and tensions to these debates (see Fig. 1).

3.1 Biodiversity versus competing land use

Converting land to agricultural use can have widespread consequences for ecosystems across scales, including loss of wildlife habitat, loss of landscape connectivity, and pollution from agricultural run-off (Rockström et al., 2017). These trade-offs between agricultural production and biodiversity are already well-described in the literature (e.g. Power, 2010; Turner et al., 2014). This research has led to calls for land-use planning approaches that integrate multiple goals (e.g. increased agricultural profits and habitat conservation) and that leverage co-benefits – unintended, beneficial outcomes from specific strategies (Karlsson et al., 2020; Newell et al., 2022). SI, as one such integrative 'co-benefit' approach, aims to use technology to farm more intensively, thereby increasing farmer profits and sparing land for biodiversity conservation. Such an approach reaffirms commitment to addressing human needs both now and, in the future, while staying within planetary and local limits.

Conversely, literature that critiques land-sparing approaches argues that increasing yields on agricultural land can have negative effects on the species living on that farmland (Phalan et al., 2014). Others instead promote land-sharing approaches that integrate biodiversity within agroecosystems (e.g. Gliessman, 2012). These debates between land-sharing versus land-sparing are already well-trodden, and indeed, the sharing versus sparing framework is increasingly questioned as a useful dichotomy (see Kremen, 2015).

Yet, novel approaches to SI are forcing us to reconsider the potential for land-sparing in more intense ways. Technologies such as controlled environment agriculture and cellular agriculture have widely promised land-saving benefits (Barbosa et al., 2015; Mattick, 2018). These technologies may spare land and take land agricultural use out of production (see, e.g. Alexander et al., 2017). There is increasing discussion, however, regarding the subsequent uses for land that is repurposed, such as whether it should be used for other forms of agricultural production, kept for conservation, or taken out of production entirely. It is not guaranteed that land would be kept 'wild' or 'naturalized' (e.g. Newman et al., 2021). These challenges are especially pertinent considering the fact that the world faces a $15 trillion infrastructure gap by 2040, with efforts being made from local to international scales to bridge this gap (WEF, 2019). As a worst case, liberated or spared agricultural land may even be zoned for urban or industrial use that all but guarantees zero biodiversity-related benefits (Newell and Glaros, forthcoming).

Were land to be converted into national parks or conservation areas, there may indeed be social and economic trade-offs as well. Fortress conservation models have well-documented negative consequences for smallholder farmers, who may be excluded from production and forced to find an alternative livelihood (Rai et al., 2021). In the case of widespread land-sparing, it is uncertain what would happen to the existing users of that land including farmers, local resource users, and communities. The ethical tension related to SI as a land-sparing approach is clear. What would happen to spared land is not! Who would have or not have access to it? How will widespread land-sparing approaches intersect with competing land-use priorities set at local levels?

3.2 Protein intensification versus human-animal relations

There is increasing global demand for livestock proteins (Henchion et al., 2017). The FAO June 2022 Food Outlook forecast that global meat production will reach 361 million tonnes (carcass weight equivalent) by the end of 2022: an increase of 1.4% since 2021 (FAO, 2022). Yet, livestock production contributes in significant ways to global environmental change (Hong et al., 2021; Cheng et al., 2022). There are thus many calls to intensify livestock production, in part achieved through increased stocking densities, concentrated feedlots, and the application of a range of technological/pharmaceutical interventions to accommodate the increased densities (Thornton, 2010).

The intensification of livestock has increased protein production by incredible volumes, but with several, severe externalities, such as decreased animal welfare and the spread of disease (Weis, 2013; Fraser, 2008). Livestock production remains a major source of carbon (methane, a potent GHG) emissions and contributes to nutrient runoff at local and regional scales (Weis, 2013). Intensified factory conditions also contribute to poor animal health, including the spread of zoonotic diseases (Gilbert et al., 2021). Antibiotic use in livestock agriculture promotes growth (and treats disease from intensified living conditions) but leads to the spread of microbial antibiotic resistance (Mann et al., 2021). This is not only a concern for animals, but for public health more broadly (Jones et al., 2013). Against this backdrop, the SI of livestock production aims to balance concerns regarding animal welfare with the need to intensify protein production using fewer land, energy, and water resources. 'New' protein markets are emerging, concentrated around the expansion and scaling of foods such as seaweed and insects that use fewer resources per kilogram of protein produced than conventional livestock (Van Huis, 2013; Smetana et al., 2016). Novel potential proteins from precision fermentation and cellular agriculture are rapidly commercializing, requiring minimal or even no use of live animals to create protein.

These novel approaches to SI of protein production generate multiple questions and trade-offs with respect to animal welfare and livestock. For example, widespread intensification would imply major economic disruptions across the livestock sector (as described by Newton and Blaustein-Rejito, 2021). What will happen to individuals engaged in the sector if their livelihoods are no longer viable? Further, novel proteins (e.g. cellular protein) may affect the way humans relate to their food, in part by adopting a 'reductive and utilitarian perspective on animal bodies' (Helliwell and Burton, 2021, p. 183). These technologies may stand to further disconnect consumers who are already distanced (physically and emotionally) from what they eat. Regarding insects, there are significant gaps in research and knowledge regarding insect-rearing welfare, and uncertainties over fundamental questions pertaining to their consciousness, their social and behavioural characteristics, and their ability to suffer from or feel pain/pleasure. These concerns have prompted recommendations to pause the widespread industrial production of insects for human or livestock consumption prior to developing ethical standards of care (Pali-Schöll et al., 2019). An ethical tension presents itself: at what social and economic cost should increased demand for sustainably produced protein be satisfied? Are alternative strategies, including reducing global demand for protein itself, more ethically justified? If reduced demand is more ethically justified, how can this strategy also be attentive to the cultural importance of various types of protein consumption across human experiences?

3.3 Quantity versus quality of food

SI aims to ensure that there are adequate supplies of food, globally (Tilman et al., 2011). Global agricultural yields have increased significantly since the early twentieth century, yet there are some signs of slowing yield growth and potential plateauing (Grassini et al., 2013). Studies have projected risk for food security, considering the following key factors: food demand, food supply, trade, productivity increases through intensification, climate change, and land availability (van Dijk et al., 2021). Noticeably absent in this list of key factors are metrics associated with human nutrition needs.

In the context of global environmental change, the pandemic, and increasingly due to geopolitical events there is a shift to broaden this narrow focus on food production, intensification, and availability. Food security includes four key dimensions: availability (supply of nutritious food), access (ability to source food), utilization (food is safe and consumable), and stability (that food is consistently available, accessible, and usable), though this definition continues to be updated (see Clapp et al., 2022). Importantly, a myriad of complex socio-political (e.g. class, race, and gender) and geo-economic factors affect an individual's ability to access and utilize high-quality food. Without ensuring its other dimensions, SI may fail to realize its food security goals. For example, substantial changes in the global yield composition are required as too few vegetables, and proteins are currently produced for adequate global nutrition (KC et al., 2018). Further, without considering if individuals can physically or financially access food, SI will not address hunger in practice. Some strategies, such as leveraging indigenous and more culturally relevant and highly nutritious crops (rather than industrially produced crops such as corn, soy, and wheat) and pursuing SI research to increase their yield would increase food availability and also utilization (Mabhaudhi et al., 2019).

On the one hand, there is an ethical tension in selecting just one or a few specific food security goals to pursue. For instance, critics of SI argue that it too narrowly focuses on production and supply dimensions. Proponents of SI instead argue that basic food needs (achieved through grain imports and availability) ought to be met before other food security goals can be realized. As Grassini and Cassman (2020) argue (p. 264):

> [a focus on nutrition] may be relevant for specific situations and local conditions, its global relevance can be questioned because it fails to recognize the reasons why the current human food supply relies so heavily on a small number of crops, and the fact that much of the developing world depends on grain imports to meet basic food needs.

There is an ethical tension between assuring all basic global nutritional needs are met now, versus pursuing holistic, aspirational measures of food security for the longer term. This tension factors into a broader ethical dilemma, recognizing that the current food system reproduces uneven power relations. How might basic food needs be met while addressing systemic barriers (e.g. classism, racism, and colonialism) that undermine the sovereignty of individuals to produce, access, and utilize high-quality foods?

3.4 Using land efficiently versus vibrant rural communities

In many parts of the world, the agricultural sector is the economic bedrock of rural communities. Approximately one billion of the world's poorest people live in rural areas and are engaged in some form of agricultural activities (Suttie, 2020). For instance, in sub-Saharan Africa the sector supports the livelihoods of 80% of the region's population, employing about 70% of the poorest on the continent as well as around 60% of the economically active population and contributes about 25% of the gross domestic products (GDP) of the region

(AGRA, 2021). Agricultural development initiatives, including the extension of technologies and capital, have intensified production and increased farmer yields globally (Juma, 2011). Suites of tools under the 'digital agriculture revolution' are emerging with the promise of further increasing farmer yield at lower environmental costs (King, 2017; Duncan et al., 2021).

Changes in yield and productivity through intensification have come at a social and economic cost for rural communities. Policies aimed at SI and sustainable development more broadly often fail to produce the desired change. For example, policy efforts to increase farmer yields and livelihoods often fail to be realized due to a lack of appropriate capital, inputs, and incentives (Adegbite and Machethe, 2020; Vanlauwe et al., 2014). Siloed sustainable development policies and programmes, often set at the national level, fail to generate whole-systems change and realize long-term effects for smallholders (Ates et al., 2018). Without comprehensive rural revitalization strategies, efforts to increase farmer yield and profit can exacerbate inequalities within communities, leading to perverse social outcomes (Kerr, 2012), a situation often described as 'deagrarianization and depeasantization'[1] (Bryceson, 1996). Access to capital (e.g. land, technology, financing) often exacerbates land consolidation, capitalization, and financialization (Duncan et al., 2022). This can manifest along class and gendered lines, where wealthier men typically have privileged access to tools and technologies to sustainably intensify production (e.g. Moseley, 2005; Agrawal, 2018). Historically, this dynamic has increased out-migration from rural communities and furthered inequalities within those communities (Mendola, 2011), continuing to the present day with digital technologies (see, e.g. Rotz et al., 2019).

There are clear ethical tensions related to the labour and livelihood consequences of SI. The first relates to the need to increase smallholder yields and incomes, while distributing the benefits of SI programmes and policies more equitably. While gains in yield and profit are accrued through intensification, there are clear and adverse implications if those benefits are not distributed more evenly within and between communities. Sustainable development and environmental governance, more broadly, requires novel assemblages of the private sector and governmental and public actors (Lemos and Agrawal, 2006). These actors work to set, implement, and assess the effects of various initiatives, including those under the umbrella of SI. However, the intended beneficiaries of such efforts (smallholders) often lack the power and capacity to advocate for themselves (Florini and Pauli, 2018).

[1] Deagrarianization is defined as a long-term process of occupational adjustment, income-earning reorientation, social identification, and spatial relocation of rural dwellers away from strictly agricultural-based modes of livelihood. Depeasantization has now begun, representing a specific form of deagrarianization in which peasantries lose their economic capacity and social coherence, and shrink in demographic size relative to non-peasant populations (Bryceson, 1996, p. 726).

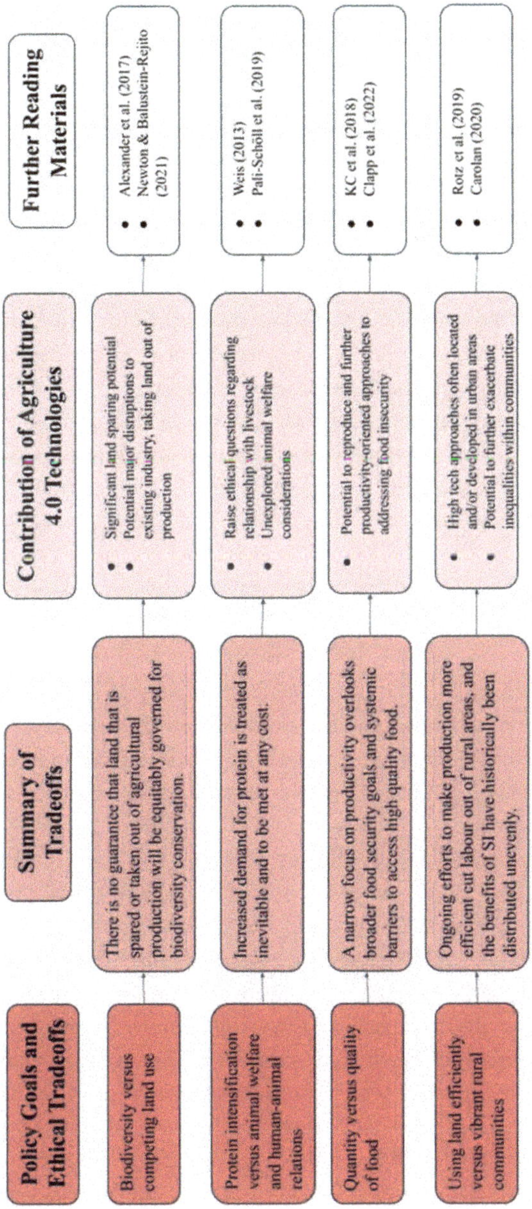

Figure 1 Summary of ethical trade-offs to SI and how agriculture 4.0 technologies contribute to these debates.

The second tension relates to the broader imperative for using land 'more efficiently'. SI, among other efforts to increase productivity per unit of area, displaces labour, thereby undermining agricultural employment in rural areas. Historically, this removal of labour from fields has resulted in migration from rural to urban areas, undermining the vibrancy and resiliency of rural communities (Lawrence, 2017). Concomitant agricultural labour shortages, exacerbated during the pandemic, drive systems of agricultural employment that rely on cheap, temporary, and precarious labour (Weiler and Grez, 2022). The growing number of jobs associated with SI and the 'digital agriculture revolution', while skilled and 'high quality' have an uneven distribution across space (Klerkx and Rose, 2020). Digital tools, among other emerging technologies for SI, are often managed, developed, and extended from urban to rural areas. Further, digital technologies are radically affecting labour politics in the countryside. The replacement of precarious agricultural labour with digital tools has been done with little to no consultation with those workers most affected (Carolan, 2020). Under SI, how might land be used efficiently to increase production while simultaneously empowering farmers and farm labour, ensuring sustainable development, and enabling vibrant rural communities?

4 Pragmatic responses to intensification-related tensions

Extrapolating from Neufeldt et al. (2013), we must hold ourselves accountable to practices in agriculture that can be shown to bring us closer to 'safe operating spaces' for agricultural and food systems across spatial and temporal scales (Neufeldt et al. 2013). In this chapter, we reflect on specific strategies and emerging technologies of agriculture 4.0 that are advancing some of the same, new, and prescient questions regarding SI and its ethical tensions. Overall, this chapter has revealed a series of ethical trade-offs that emerge when trying to apply SI to guide real-world agri-food system management. Namely, a single-minded SI approach might result in intensively farmed land but no benefits to biodiversity; increased production of grains and oilseeds but no commensurate increase in other food commodities that are culturally appropriate to some and needed for a healthful diet; increased concentrations of animal agriculture; and less vibrant rural communities. To ensure that we avoid these suboptimal outcomes, each trade-off must be approached pragmatically considering all potential institutional, governance-related, or technological solutions at hand. In this section, we outline why a pragmatic approach to SI is needed and how it may be realized in practice. A pragmatic approach to SI would be multi-pronged, evaluating actions for sustainable production on a case-by-case basis (Tilman et al., 2011; Godfray and Garnett,

2015). Furthermore, a pragmatic SI approach rejects the assertion that any one strategy to change our food systems is sufficient (e.g. Fraser et al., 2016; Fanzo et al., 2021). A single approach to changing food systems could be very detrimental. For example, the Sri Lankan government's abrupt transition of the country's agriculture systems to organic farming in 2021 resulted in poor crop yields and productivity, revenue losses for farmers, high food prices, and a surge in the country's already growing economic crises; this ultimately led to the ousting of the president through civil unrest.[2] Finally, a pragmatic approach ought to recognize the need and advocate for the pursuit of more inclusive SI governance mechanisms, to ensure all benefits of SI are more evenly distributed. In this discussion, we review each of the ethical tensions described in Section 3 and delineate pragmatic responses to those issues (Table 1).

Proponents of SI argue against a priori responses to sustainability-related challenges. Thus, as Garnett et al. (2013) suggest, land-sharing and/or land-sparing approaches may make sense in diverse contexts, on a case-by-case basis. Importantly, any decision as to which type of approach ought to be taken needs to be supported by appropriate, high-quality data. Further, if emerging technologies for SI do spare land, there is a need for further research on what potential uses of that land could be repurposed to achieve the most social, economic, and environmental goals.

To address the challenge of satisfying increased protein without compromising animal welfare, SI responses advocate for the adoption of a wider ethical framework. In this case, one ethical contention is that all humans should have the opportunity to access any amount of desired livestock protein at a minimal environmental cost, while another is that the dietary preferences of humans should not outweigh the welfare of animals (or industry workers). A pragmatic approach to SI ought to negotiate between these diverse contentions and challenge the assumptions underlying each imperative (i.e. that there should or will be increased demand for animal protein). More broadly, a pragmatic approach would recognize, validate, and work between multiple ethical contentions.

Another key tension regarding SI is its need to balance diverse food security dimensions. Scholars recommend that a more comprehensive definition of 'food security' be used to develop SI initiatives that consider multiple dimensions including the nutritional quality of food (e.g. KC et al., 2018), its utilization, accessibility, and stability over time. There are many emerging examples of deliberative, inclusive governance mechanisms that capture multiple dimensions of food security. For example, some research

[2] Sri Lanka saw rice and tea yields fall 30% and 18%, respectively, upon their abrupt ban of agrochemical imports in spring of 2021. https://time.com/6196570/sri-lanka-crisis-organic-farming/.

Table 1 Tensions, solutions, and pragmatic roots for SI

Tensions	Pragmatic solutions	Illustrative examples and case studies
Yield increases and farmer livelihood at odds with conservation-oriented land-use policies	• Increase available empirical evidence/data required to evaluate if land-sharing or land-sparing makes sense and how to best implement it, in a given biophysical and/or socio-political context (Garnett, 2014)	• Evaluate land-use scenarios and policies under different land-sparing scenarios, considering emerging SI approaches and/or technologies (e.g. Newman et al., 2021)
Satisfying increased demand for protein without sacrificing animal well-being in intensified systems	• SI placed within wider ethical framework to consider animal welfare, human-animal relations, and consequences on livestock industry workers • Consider alternative approaches, including those that contest the premise that protein demand will or must increase (Garnett et al., 2014) • Conceptualize and design solutions further downstream from livestock production (e.g. curbing livestock protein demand)	• Intensify production without increasing stocking densities through, for example, monitoring for optimal health of stock, embedding protein production in circular economic streams (Taifouris and Martin, 2021) • Continued research and development of cellular agriculture and alternative plant-based protein sources
Pursuing all dimensions of food security, while assuring that basic food needs are met in the short-term	• Develop comprehensive measures for food security • Ensuring that global food production meets the macro- and micro-nutrient needs of the population and address areas of underproduction • Evaluate nutritional properties of intensified production and take inventory of diverse food needs from local communities • Rank and evaluate actionable food security priorities	• Pursue SI interventions for culturally relevant foods (i.e. not corn, soy, maize) (Mabhaudhi et al., 2019)

(*Continued*)

Table 1 (*Continued*)

Tensions	Pragmatic solutions	Illustrative examples and case studies
Balancing broader sustainable development priorities with local contexts and needs	• Develop sustainable development policies with needs of smallholders in mind • Include smallholders in design of sustainable development policies and practices	• Engage in responsible innovation and research processes for SI approaches, particularly those that leverage emerging agriculture 4.0 technologies (e.g. Wittman et al., 2020) • Strengthen seed systems to enhance farmers access to improved seeds and complementing technologies (Quarshie et al., 2021, 2022) • Include more accessible location and context-specific extension programmes for smallholders (Quarshie et al., 2022; Abdulai, forthcoming)
Increasing farmer livelihoods and profits without exacerbating inequalities within rural communities	• Pursue pro-poor policies and investments to stimulate the sector's growth as part of a broader focus on agriculture and rural development (Giller et al., 2021) • Continuously monitor policy outcomes on rural communities and livelihoods • Pursue a rural innovation system which aims to revitalize rural areas to achieve balanced sustainable development (Yin et al., 2019) • Incentivize participation of smallholders without existing access to capital	• Re-evaluate outcomes of policy interventions continuously to determine their efficacy • Co-develop extension services (e.g. Pretty et al., 2018) • Expand rural non-farm economic opportunities for rural livelihood diversification (Haggblade et al., 2010; Dzanku et al., 2015)

is cataloguing and exploring the dietary diversity of various communities (e.g. Powell et al., 2017). Alongside local farmers, these findings could be leveraged to create high-yielding, nutritional foods of cultural relevance. However, other scholars suggest approaches such as these do not address the globe's most prescient, basic caloric needs. Here, the priority is getting basic food commodities to those who most need them. A pragmatic approach to negotiating these tensions would involve potentially ranking diverse priorities for SI (see, e.g. Eppinga et al., 2022).

Further, a pragmatic approach to SI would pursue a democratic, deliberative approach to decision-making. This is particularly relevant for the challenge that top-down sustainable development policymaking has failed in many cases to accommodate local knowledge and needs. Including smallholder needs and input into sustainable development is a more democratic and reflexive approach. Likewise, the involvement of indigenous peoples in decision-making on their livelihoods or food systems upholds their right to self-determination as enshrined within the United Nations Declaration on the Rights of Indigenous Peoples (UNDRIP) (UNGA, 2007). To pursue this, for example, some scholars recommend the formation of data cooperatives as a means to safeguard farmer-derived field knowledge (Cassman and Grassini, 2020). Other scholars promote the co-production of environmental knowledge, using novel extension approaches that better leverage local knowledge alongside scientific approaches to yield improvement (Pretty et al., 2018). Balancing broadly defined sustainable development goals with local knowledge and context would engender more democratic SI outcomes.

Similarly, concerns over the implementation of SI and its effects on rural communities have highlighted a variety of pragmatic proposals to ensure SI benefits are more evenly distributed. Inclusive governance mechanisms are crucial for SI policy and programme to realize these widespread benefits, particularly for the world's poorest farmers who also happen to be the most food insecure. Pragmatists approach sustainability problems with the goal of negotiating, compromising, and including all relevant stakeholders to arrive at 'best-case' scenarios. As previously noted, historical approaches to intensification and SI have generated uneven development, prioritizing those with existing access to capital and reinforcing divisions across class and gendered lines (Kerr, 2012). Emerging technologies of agriculture 4.0 may further those same inequalities (see Fig. 1). A pragmatic SI approach would include these considerations and develop participatory mechanisms from the outset of programme development, continuously evaluating and re-negotiating programmes based on their outcomes.

One core philosophical concept that helps unify these strategies is 'environmental pragmatism' borne from early thinkers such as John Dewey who critiqued what they considered philosophy's abstract approach to analysing morals and ethics in unambiguous, unchanging terms. Instead, the pragmatists viewed morals and ethics as continuously evolving categories, embedded in society, evolving toward 'better' but imperfect futures (Katz and Light, 1996). Early environmental pragmatists scrutinized existing philosophical debates regarding the environment through this lens. These debates largely sought to justify nature's 'worth' on the basis of intrinsic (rather than instrumental, i.e. economic) value (e.g. Callicott, 1995).

Environmental pragmatists argued that the value of nature cannot be divorced from practical goals for conservation or the culture in which conservation policies are embedded (Katz and Light, 1996). Instead, environmental activism and policy are inherently embedded in social and political systems. Philosophers arrive at the 'truth' to various philosophical questions (e.g. what is nature's worth?) not through philosophizing at a desk but through doing environmental work on the ground. Through continuous negotiation, scientific discovery, and inquiry, society arrives at better but imperfect understandings of nature. As Dr. Bryan Norton, a pioneer of environmental pragmatism describes:

> Pragmatism is a forward-looking philosophy. The truth is that Emerges over time. We're always going to make mistakes, but science, through many, many iterations, gradually gets further away from error and maybe even a little closer to the truth. So that's the pragmatist view that we'll never have complete knowledge because it's constantly changing (Interview with Dr. Bryan Norton, 2016).

We conclude by highlighting what we believe to be the three key tenets of environmental pragmatism that apply to SI. The first tenet is that pragmatists reject dichotomies which, they argue, are overly abstract and ineffectual when it comes to policymaking and environmental action (Tuminello, 2019). More specifically, they reject moral monism, the idea that a single moral or ethical framework can be applied to every situation. Here, pragmatists argue for practical, case-by-case action to achieve environmental goals, rather than applying pre-determined moral or ethical frameworks to a situation. In the case of SI, this would require the rejection of a priori assumptions for what approach to bettering our food system should be taken. This is particularly relevant for emerging technologies for SI, where in some contexts these tools will be invaluable, while in others they may be unhelpful or even problematic. Further, this would require a stronger emphasis on doing rather than theorizing sustainability. This requires a commitment, particularly on the part of researchers, to undertake an empirically and methodologically rigorous collection of quantitative and qualitative data to evaluate policies and programmes in real-time.

The second tenet is as follows: nature is given value intrinsically or instrumentally which leads to the same form of conservation policies or actions (Norton, 1984). In other words, pragmatists seek common ground over shared goals, rather than finding division between 'competing' approaches, strategies, or ideologies. SI is clearly a contested term (see, e.g. Struik and Kuyper, 2017). Pragmatism requires parties for or against the term to establish common ground over shared goals, such as those defined in the previous section: increasing global biodiversity, addressing animal welfare concerns, meeting all measures for global food security, and supporting rural communities. The key here is ensuring that all relevant parties with shared goals, regardless of their positionality, are included and have equal capacity to participate in decision-making.

The third tenet is pragmatists advocate for a democratic and deliberative approach to environmental decision-making (Overdevest et al., 2010). Such approaches better account for 'surprises', unexpected outcomes, and diverse stakeholder needs in the pursuit of environmental goals. This applies to both environmental decision-making and agri-food research design. Instruction can come here from emerging best practices in research, including community-based approaches that match research capacity with partner needs. For agri-food innovation development, this may require the adoption of responsible innovation practices that incorporate farmers into the design, development, and on-going use of technologies for food production (see, e.g. Wittman et al., 2020).

5 Conclusion

This chapter presents readers with a review of the debates and ethical tensions - new and old - regarding SI, considering the novel technologies of agriculture 4.0. It argues that environmental pragmatism as a philosophical approach to policymaking and research is a useful approach to negotiating those tensions. The subject matter we review is extensive, and we thus present Table 2 to summarize the key debates and present readers with a list of relevant reading materials to consult for further exploration.

6 Where to look for further information

In this chapter, we have highlighted the underlying premises and importance of SI. We have described key policy goals for SI, illuminating the underlying tensions that make it a contested and often fraught approach to food systems' transformation. We have also reflected on novel agricultural technologies that force us to reconsider the potential ethical trade-offs underpinning SI. We argue that a pragmatic approach is needed for its governance and implementation in practice. Indeed, SI has many commonalities with environmental pragmatism as a philosophical orientation to the pursuit of sustainability-related goals. We conclude by sketching a number of pragmatic responses to the ethical tensions we previously identified. In so doing, we recommend that SI approaches

Table 2 Relevant articles for further exploration

Key topics	Relevant articles
Sustainable intensification debates	• Cassman and Grassini (2020). A global perspective on sustainable intensification research. *Nature Sustainability*, *3*(4), 262–268. https://doi.org/10.1038/s41893-020-0507-8 • Garnett et al. (2013). Sustainable Intensification in Agriculture: Premises and Policies. *Science*, *341*(6141), 33–34. https://doi.org/10.1126/science.1234485 • Struik and Kuyper (2017). Sustainable intensification in agriculture: the richer shade of green. A review. *Agronomy for Sustainable Development*, *37*(5), 39. https://doi.org/10.1007/s13593-017-0445
Agriculture 4.0 technology tradeoffs	• Bierbaum et al. (2020). Novel entities and technologies: Environmental benefits and risks. *Environmental Science & Policy*, *105*, 134–143. https://doi.org/10.1016/j.envsci.2019.11.002 • Glaros et al. (2021). Horizon scanning and review of the impact of five food and food production models for the global food system in 2050. *Trends in Food Science & Technology*, *119*, 550–564. https://doi.org/10.1016/j.tifs.2021.11.013 • Klerkx and Rose (2020). Dealing with the game-changing technologies of Agriculture 4.0: How do we manage diversity and responsibility in food system transition pathways? *Global Food Security*, *24*, 100347. https://doi.org/10.1016/j.gfs.2019.100347
Environmental pragmatism	• Katz and Light (1996). *Environmental Pragmatism*. Routledge

maintain an emphasis on democratic deliberation and inclusive as well as reflexive governance processes. Further, we caution against dualisms and arguments for food systems' transformation that do not adequately consider local context or available data.

SI is but one approach to achieving 'better' food futures. It aims to reconcile diverse perspectives on food systems' change, generating compromise, action, and (hopefully) positive results across scales. By defining its roots in pragmatism, we can see how it differs from strategies to food system change with a priori assumptions for what strategies ought to be used and how change ought to occur. Pragmatic approaches such as SI take a long, complex, and arduous road to arrive at 'the truth' or 'positive' outcomes. With compromise, this approach will never be perfect. Moreover, it requires us to observe progress not necessarily in the here and now but in the longue durée. However, its cautious, data-driven implementation can steer society and communities toward hopeful futures, where perhaps agriculture and food systems do not undermine the economic, social, and environmental bases that generate food security and nutrition for present and future generations.

7 References

Abdulai, A. R. Are rural smallholders ready for agricultural digitalisation? (In)competencies and (the lack of) access in digital agricultural extension and advisories (AEAS) in Northern Ghana. In *Digital Communication for Agricultural and Rural Development*. Taylor and Francis: Oxfordshire.

Adegbite, O. O. and Machethe, C. L. (2020). Bridging the financial inclusion gender gap in smallholder agriculture in Nigeria: An untapped potential for sustainable development. *World Development* 127, 104755. https://doi.org/10.1016/j.worlddev.2019.104755.

AGRA (2021). *Africa's Agriculture Status Report 2021*. Nairobi, Kenya: Alliance for a Green Revolution in Africa (AGRA). Available at: https://agra.org/wp-content/uploads/2021/09/AASR2021-A-Decade-of-Action-_Building-Sustainable-and-Resilient-Food-Systems-in-Africa.pdf.

Agrawal, B. (2018). Gender equality, food security and the sustainable development goals. *Current Opinion in Environmental Sustainability* 34, 26–32.

Alexander, P., Brown, C., Arneth, A., Dias, C., Finnigan, J., Moran, D. and Rounsevell, M. D. A. (2017). Could consumption of insects, cultured meat or imitation meat reduce global agricultural land use? *Global Food Security* 15, 22–32. https://doi.org/10.1016/j.gfs.2017.04.001.

Altieri, M. and Nicholls, C. (2012). Agroecology scaling up for food sovereignty and resiliency. In: *Sustainable Agriculture Research* 11. https://doi.org/10.1007/978-94-007-5449-2_1.

Ates, S., Cicek, H., Bell, L. W., Norman, H. C., Mayberry, D. E., Kassam, S., Hannaway, D. B. and Louhaichi, M. (2018). Sustainable development of smallholder crop-livestock farming in developing countries. *IOP Conference Series: Earth and Environmental Science* 142, 012076. https://doi.org/10.1088/1755-1315/142/1/012076.

Barbosa, G. L., Gadelha, F. D. A., Kublik, N., Proctor, A., Reichelm, L., Weissinger, E., Wohlleb, G. M. and Halden, R. U. (2015). Comparison of land, water, and energy requirements of lettuce grown using hydroponic vs. conventional agricultural methods. *International Journal of Environmental Research and Public Health* 12(6), 6879–6891. https://doi.org/10.3390/ijerph120606879.

Bernard, B. and Lux, A. (2017). How to feed the world sustainably: an overview of the discourse on agroecology and sustainable intensification. *Regional Environmental Change* 17(5), 1279–1290. https://doi.org/10.1007/s10113-016-1027-y.

Bierbaum, R., Leonard, S. A., Rejeski, D., Whaley, C., Barra, R. O. and Libre, C. (2020). Novel entities and technologies: environmental benefits and risks. *Environmental Science and Policy* 105, 134–143. https://doi.org/10.1016/j.envsci.2019.11.002.

Bryceson, D. F. (1996). Deagrarianization and rural employment in sub-Saharan Africa: A sectoral perspective. *World Development* 24(1), 97–111.

Callicott, J. B. (1995). Intrinsic value in nature: A metaethical analysis. *Electronic Journal of Analytic Philosophy* 3(5).

Campbell, B. M., Thornton, P., Zougmoré, R., van Asten, P. and Lipper, L. (2014). Sustainable intensification: What is its role in climate smart agriculture? *Current Opinion in Environmental Sustainability* 8, 39–43. https://doi.org/10.1016/j.cosust.2014.07.002.

Carolan, M. (2020). Automated agrifood futures: Robotics, labor and the distributive politics of digital agriculture. *The Journal of Peasant Studies* 47(1), 184-207. https://doi.org/10.1080/03066150.2019.1584189.

Cassman, K. G. and Grassini, P. (2020). A global perspective on sustainable intensification research. *Nature Sustainability* 3(4), 262–268. https://doi.org/10.1038/s41893-020-0507-8.

Cheng, M., McCarl, B. and Fei, C. (2022). Climate change and livestock production: A literature review. *Atmosphere* 13(1), 140. https://doi.org/10.3390/atmos13010140.

Clapp, J.,Moseley, W. G., Burlingame, B. and Termine, P. (2022). Viewpoint: The case for a six dimensional food security framework. *Food Policy* 106, 102164. https://doi.org/10.1016/j.foodpol.2021.102164.

Collins, E. D. and Chandrasekaran, K. (2012). *A Wolf in Sheep's Clothing? An Analysis of the 'Sustainable Intensification' of Agriculture*. Amsterdam, The Netherlands: Friends of the Earth International.

Conway, G. (2012). *One Billion Hungry, Can We Feed the World?* Ithaca: Cornell University Press.

Donovan, M. (2020). What is sustainable intensification? Farming method can boost yields, increase farmer's profits and reduce greenhouse gas emissions. *CIMMYT*. Accessed April 1, 2022. Available at: https://www.cimmyt.org/news/what-is-sustainable-intensification/.

Duncan, E., Abdulai, A. R. and Fraser, E. D. (2021). *Modernizing Agriculture Through Digital Technologies: Prospects and Challenges. Handbook on the Human Impact of Agriculture* (pp. 138–161).

Duncan, E., Rotz, S., Magnan, A. and Bronson, K. (2022). Disciplining land through data: The role of agricultural technologies in farmland assetisation. *Sociologia Ruralis* 62(2), 231-249.

Dzanku, F. M., Jirström, M. and Marstorp, H. (2015). Yield gap-based poverty gaps in rural SubSaharan Africa. *World Development* 67, 336–362.

Eppinga, M. B., Mijts, E. N. and Santos, M. J. (2022). Ranking the sustainable development goals: Perceived sustainability priorities in small island states. *Sustainability Science* 17(4), 1537–1556. https://doi.org/10.1007/s11625-022-01100-7.

Evans, A. (2011). *Resource Scarcity, Climate Change and the Risk of Violent Conflict*. Center for International Cooperation (p. 24) [World Development Report 11: Background Paper].

Fanzo, J., Haddad, L., Schneider, K. R., Béné, C., Covic, N. M., Guarin, A., Herforth, A. W., Herrero, M., Sumaila, U. R., Aburto, N. J., Amuyunzu-Nyamongo, M., Barquera, S., Battersby, J., Beal, T., Bizzotto Molina, P., Brusset, E., Cafiero, C., Campeau, C., Caron, P., Cattaneo, A., Conforti, P., Davis, C., DeClerck, F. A. J., Elouafi, I., Fabi, C., Gephart, J. A., Golden, C. D., Hendriks, S. L., Huang, J., Laar, A., Lal, R., Lidder, P., Loken, B., Marshall, Q., Masuda, Y. J., McLaren, R., Neufeld, L. M., Nordhagen, S., Remans, R., Resnick, D., Silverberg, M., Torero Cullen, M., Tubiello, F. N., Vivero-Pol, J., Wei, S. and Rosero Moncayo, J. (2021). Viewpoint: Rigorous monitoring is necessary to guide food system transformation in the countdown to the 2030 global goals. *Food Policy* 104, 102163. https://doi.org/10.1016/j.foodpol.2021.102163.

FAO (2016). Scaling-up the sustainable intensification of agriculture. Accessed April 1, 2022. Available at: https://www.fao.org/3/i5653e/i5653e.pdf.

FAO (2022). *Food Outlook – Biannual Report on Global Food Markets*. Rome: FAO. https://doi.org/10.4060/cb9427en.

Florini, A. and Pauli, M. (2018). Collaborative governance for the Sustainable Development Goals. *Asia and the Pacific Policy Studies* 5(3), 583–598. https://doi.org/10.1002/app5.252.

Fraser, D. (2008). Understanding animal welfare. *Acta Veterinaria Scandinavica* 50(Suppl. 1), S1-12. https://doi.org/10.1186/1751-0147-50-S1-S1.

Fraser, E., Legwegoh, A. and Kc, K. (2016). Biotechnology or organic? Extensive or intensive? Global or local? A critical review of potential pathways to resolve the global food crisis. *Trends in Food Science & Technology* 48, 78–87. https://doi.org/10.1016/j.tifs.2015.11.006.

Fraser, E. D. G. and Campbell, M. (2019). Agriculture 5.0: Reconciling production with planetary health. *One Earth* 1(3), 278–280. https://doi.org/10.1016/j.oneear.2019.10.022.

Garnett, T. (2013). Food sustainability: Problems, perspectives and solutions. *Proceedings of the Nutrition Society* 72(1), 29–39. https://doi.org/10.1017/S0029665112002947.

Garnett, T. (2014). Restraint, food system transformation. *Journal of Cleaner Production* 73, 10–18. https://doi.org/10.1016/j.jclepro.2013.07.045.

Garnett, T., Appleby, M. C., Balmford, A., Bateman, I. J., Benton, T. G., Bloomer, P., Burlingame, B., Dawkins, M., Dolan, L., Fraser, D., Herrero, M., Hoffmann, I., Smith, P., Thornton, P. K., Toulmin, C., Vermeulen, S. J. and Godfray, H. C. J. (2013). Agriculture: Sustainable intensification in agriculture: premises and policies. *Science* 341(6141), 33–34. https://doi.org/10.1126/science.1234485.

Gilbert, W., Thomas, L. F., Coyne, L. and Rushton, J. (2021). Review: mitigating the risks posed by intensification in livestock production: The examples of antimicrobial resistance and zoonoses. *Animal: An International Journal of Animal Bioscience* 15(2), 100123. https://doi.org/10.1016/j.animal.2020.100123.

Giller, K. E., Delaune, T., Silva, J. V., Descheemaeker, K., van de Ven, G., Schut, A. G. T., van Wijk, M., Hammond, J., Hochman, Z., Taulya, G., Chikowo, R., Narayanan, S., Kishore, A., Bresciani, F., Teixeira, H. M., Andersson, J. A. and van Ittersum, M. K. (2021). The future of farming: Who will produce our food? *Food Security* 13(5), 1073–1099. https://doi.org/10.1007/s12571-021-01184-6.

Glaros, A., Marquis, S., Major, C., Quarshie, P., Ashton, L., Green, A. G., Kc, K. B., Newman, L., Newell, R., Yada, R. Y. and Fraser, E. D. G. (2022). Horizon scanning and review of the impact of five food and food production models for the global food system in 2050. *Trends in Food Science and Technology* 119, 550–564. https://doi.org/10.1016/j.tifs.2021.11.013.

Gliessman, S. (2012). Agroecology and shifting paradigms. *Journal of Sustainable Agriculture* 36(5), 499–499. https://doi.org/10.1080/10440046.2012.684200.

Godfray, H. C. J. (2015). The debate over sustainable intensification. *Food Security* 7(2), 199–208. https://doi.org/10.1007/s12571-015-0424-2.

Godfray, H. C. J., Beddington, J. R., Crute, I. R., Haddad, L., Lawrence, D., Muir, J. F., Pretty, J., Robinson, S., Thomas, S. M. and Toulmin, C. (2010). Food security: The challenge of Feeding 9 billion people. *Science* 327(5967), 812–818. https://doi.org/10.1126/science.1185383.

Godfray, H. C. J. and Garnett, T. (2014). Food security and sustainable intensification. *Philosophical Transactions of the Royal Society of London: Series B, Biological Sciences* 369(1639), 20120273. https://doi.org/10.1098/rstb.2012.0273.

Grassini, P., Eskridge, K. M. and Cassman, K. G. (2013). Distinguishing between yield advances and yield plateaus in historical crop production trends. *Nature Communications* 4, 2918. https://doi.org/10.1038/ncomms3918.

Green, A. G., Abdulai, A., Duncan, E., Glaros, A., Campbell, M., Newell, R., Quarshie, P., Kc, K. B., Newman, L., Nost, E. and Fraser, E. D. G. (2021). A scoping review of the digital agricultural revolution and ecosystem services: Implications for Canadian policy and research agendas. *Facets* 6, 1955–1985. https://doi.org/10.1139/facets-2021-0017.

Haggblade, S., Hazell, P. and Reardon, T. (2010). The rural non-farm economy: Prospects for growth and poverty reduction. *World Development* 38(10), 1429–1441. https://doi.org/10.1016/j.worlddev.2009.06.008.

Helliwell, R. and Burton, R. J. F. (2021). The promised land? Exploring the future visions and narrative silences of cellular agriculture in news and industry media. *Journal of Rural Studies* 84, 180–191.

Henchion, M., Hayes, M., Mullen, A. M., Fenelon, M. and Tiwari, B. (2017). Future protein supply and demand: Strategies and factors influencing a sustainable equilibrium. *Foods (Basel, Switzerland)* 6(7), E53. https://doi.org/10.3390/foods6070053.

Holt-Giménez, E., Shattuck, A., Altieri, M., Herren, H. and Gliessman, S. (2012). We already grow enough food for 10 billion people ... and still can't end hunger. *Journal of Sustainable Agriculture* 36(6), 595–598. https://doi.org/10.1080/10440046.2012.695331.

Homer-Dixon, T., Maynard, J. L., Mildenberger, M., Milkoreit, M., Mock, S. J., Quilley, S., Schröder, T. and Thagarde, P. (2013). A complex systems approach to the study of ideology: cognitive affective structures and the dynamics of belief change. *Journal of Social and Political Psychology* 1(1), 337–363. DOI: 10.5964/jspp.v1i1.36.

Hong, C., Burney, J. A., Pongratz, J., Nabel, J. E. M. S., Mueller, N. D., Jackson, R. B. and Davis, S. J. (2021). Global and regional drivers of land-use emissions in 1961–2017. *Nature* 589(7843), 554–561. https://doi.org/10.1038/s41586-020-03138-y.

Hunter, M. C., Smith, R. G., Schipanski, M. E., Atwood, L. W. and Mortensen, D. A. (2017). Agriculture in 2050: Recalibrating targets for sustainable intensification. *BioScience* 67(4), 386–391.

Interview with Dr. Bryan Norton (2016). *The Pragmatist's View: A Conversation With Bryan Norton Retrievable*. Available at: https://humansandnature.org/the-pragmatists-view-a-conversationwith-bryan-norton/.

Jones, B. A., Grace, D., Kock, R., Alonso, S., Rushton, J., Said, M. Y., McKeever, D., Mutua, F., Young, J., McDermott, J. and Pfeiffer, D. U. (2013). Zoonosis emergence linked to agricultural intensification and environmental change. *Proceedings of the National Academy of Sciences of the United States of America* (21), 8399–8404. https://doi.org/10.1073/pnas.1208059110.

Juma, C. (2011). *The New Harvest: Agricultural Innovation in Africa*. Oxford University Press, Oxford.

Karlsson, M., Alfredsson, E. and Westling, N. (2020). Climate policy co-benefits: A review. *Climate Policy* 20(3), 292–316. https://doi.org/10.1080/14693062.2020.1724070.

Katz, E. and Light, A. (1996). *Environmental Pragmatism*. Routledge, London.

Kc, K. B., Dias, G. M., Veeramani, A., Swanton, C. J., Fraser, D., Steinke, D., Lee, E., Wittman, H., Farber, J. M., Dunfield, K., McCann, K., Anand, M., Campbell, M., Rooney, N., Raine, N. E., Acker, R. V., Hanner, R., Pascoal, S., Sharif, S., Benton, T. G. and Fraser, E. D. G. (2018). When too much isn't enough: Does current food production meet global nutritional needs? *PLoS ONE* 13(10), e0205683. https://doi.org/10.1371/journal.pone.0205683.

Kerr, R. B. (2012). Lessons from the old Green Revolution for the new: Social, environmental and nutritional issues for agricultural change in Africa. *Progress in Development Studies* 12(2–3), 213–229. https://doi.org/10.1177/146499341101200308.

King, A. (2017). Technology: the future of agriculture. *Nature* 544(7651), S21–S23.

Klerkx, L. and Rose, D. (2020). Dealing with the game-changing technologies of Agriculture 4.0: How do we manage diversity and responsibility in food system transition pathways? *Global Food Security* 24, 100347. https://doi.org/10.1016/j.gfs.2019.100347.

Kremen, C. (2015). Reframing the land-sparing/land-sharing debate for biodiversity conservation. *Annals of the New York Academy of Sciences* 1355(1), 52–76. https://doi.org/10.1111/nyas.12845.

Lawrence, G. (2017). Re-evaluating food systems and food security: A global perspective. *Journal of Sociology* 53(4), 774–796. https://doi.org/10.1177/1440783317743678.

Lemos, M. C. and Agrawal, A. (2006). Environmental governance. *Annual Review of Environment and Resources* 31(1), 297–325. https://doi.org/10.1146/annurev.energy.31.042605.135621.

Mabhaudhi, T., Chibarabada, T. P., Chimonyo, V. G. P., Murugani, V. G., Pereira, L. M., Sobratee, N., Govender, L., Slotow, R. and Modi, A. T. (2019). Mainstreaming underutilized indigenous and traditional crops into food systems: A South African perspective. *Sustainability* 11(1), 172. https://doi.org/10.3390/su11010172.

Mahon, N., Crute, I., Di Bonito, M., Simmons, E. A. and Islam, M. M. (2018). Towards a broadbased and holistic framework of sustainable intensification indicators. *Land Use Policy* 77(C), 576-597. Available at: https://ideas.repec.org/a/eee/lauspo/v77y2018icp576-597.html.

Mann, A., Nehra, K., Rana, J. S. and Dahiya, T. (2021). Antibiotic resistance in agriculture: perspectives on upcoming strategies to overcome upsurge in resistance. *Current Research in Microbial Sciences* 2, 100030. https://doi.org/10.1016/j.crmicr.2021.100030.

Mattick, C. S. (2018). Cellular agriculture: The coming revolution in food production. *Bulletin of the Atomic Scientists* 74(1), 32–35. https://doi.org/10.1080/00963402.2017.1413059.

Mendola, M. (2012). Rural out-migration and economic development at origin: a review of the evidence. *Journal of International Development* 24(1), 102–122. https://doi.org/10.1002/jid.1684.

Mouratiadou, I., Latka, C., van der Hilst, F., Müller, C., Berges, R., Bodirsky, B. L., Ewert, F., Faye, B., Heckelei, T., Hoffmann, M., Lehtonen, H., Lorite, I. J., Nendel, C., Palosuo, T., Rodríguez, A., Rötter, R. P., Ruiz-Ramos, M., Stella, T., Webber, H. and Wicke, B. (2021). Quantifying sustainable intensification of agriculture: The contribution of metrics and modelling. *Ecological Indicators* 129, 107870. https://doi.org/10.1016/j.ecolind.2021.107870.

Moseley, W. G. (2005). Global cotton and local environmental management: the political ecology of rich and poor small-hold farmers in southern Mali. *The Geographical Journal* 171(1), 36–55. https://doi.org/10.1111/j.1475-4959.2005.00148.x.

Neufeldt, H., Jahn, M., Campbell, B. M., Beddington, J. R., DeClerck, F., De Pinto, A., Gulledge, J., Hellin, J., Herrero, M., Jarvis, A., LeZaks, D., Meinke, H., Rosenstock, T., Scholes, M., Scholes, R., Vermeulen, S., Wollenberg, E. and Zougmoré, R. (2013). Beyond climate-smart agriculture: Toward safe operating spaces for global food systems. *Agriculture and Food Security* 2(1), 12. https://doi.org/10.1186/2048-7010-2-12.

Newell, R., Dale, A. and Lister, N.-M. (2022). An integrated climate-biodiversity framework to improve planning and policy: An application to wildlife crossings and landscape connectivity. *Ecology and Society* 27(1). https://doi.org/10.5751/ES-12999-270123.

Newell, R. and Glaros, A. Sustainable food systems, development paths, and the futures of cellular agriculture. In: *Cellular Agriculture: Technical and Scientific Foundations*. Elsevier.

Newman, L., Newell, R., Mendly-Zambo, Z. and Powell, L. (2022). Bioengineering, telecoupling, and alternative dairy: Agricultural land use futures in the Anthropocene. *Geographical Journal* 188(3), 342–357. https://doi.org/10.1111/geoj.12392.

Newton, P. and Blaustein-Rejto, D. (2021). Social and economic opportunities and challenges of plant-based and cultured meat for rural Producers in the US. *Frontiers in Sustainable Food Systems* 5. Available at: https://www.frontiersin.org/article/10.3389/fsufs.2021.624270.

Overdevest, C., Bleicher, A. and Gross, M. (2010). The experimental turn in environmental sociology: pragmatism and new forms of governance. In: Gross, M. and Heinrichs, H. (Eds.), *Environmental Sociology: European Perspectives and Interdisciplinary Challenges* (pp. 279–294). Springer Netherlands. https://doi.org/10.1007/978-90-481-8730-0_16.

Pali-Schöll, I., Binder, R., Moens, Y., Polesny, F. and Monsó, S. (2019). Edible insects – Defining knowledge gaps in biological and ethical considerations of entomophagy. *Critical Reviews in Food Science and Nutrition* 59(17), 2760–2771. https://doi.org/10.1080/10408398.2018.1468731.

Phalan, B., Green, R. and Balmford, A. (2014). Closing yield gaps: Perils and possibilities for biodiversity conservation. *Philosophical Transactions of the Royal Society of London: Series B, Biological Sciences* 369(1639), 20120285. https://doi.org/10.1098/rstb.2012.0285.

Powell, B., Bezner Kerr, R., Young, S. L. and Johns, T. (2017). The determinants of dietary diversity and nutrition: Ethnonutrition knowledge of local people in the East Usambara Mountains, Tanzania. *Journal of Ethnobiology and Ethnomedicine* 13(1), 23. https://doi.org/10.1186/s13002-017-0150-2.

Power, A. G. (2010). Ecosystem services and agriculture: Tradeoffs and synergies. *Philosophical Transactions of the Royal Society of London. Series B, Biological Sciences* 365(1554), 2959–2971. https://doi.org/10.1098/rstb.2010.0143.

Pretty, J., Benton, T. G., Bharucha, Z. P., Dicks, L. V., Flora, C. B., Godfray, H. C. J., Goulson, D., Hartley, S., Lampkin, N., Morris, C., Pierzynski, G., Prasad, P. V. V., Reganold, J., Rockström, J., Smith, P., Thorne, P. and Wratten, S. (2018). Global assessment of agricultural system redesign for sustainable intensification. *Nature Sustainability* 1(8), 441–446. https://doi.org/10.1038/s41893-018-0114-0.

Pretty, J. and Bharucha, Z. P. (2014). Sustainable intensification in agricultural systems. *Annals of Botany* 114(8), 1571–1596. https://doi.org/10.1093/aob/mcu205.

Pretty, J., Sutherland, W. J., Ashby, J., Auburn, J., Baulcombe, D., Bell, M., Bentley, J., Bickersteth, S., Brown, K., Burke, J., Campbell, H., Chen, K., Crowley, E., Crute, I., Dobbelaere, D., Edwards-Jones, G., Funes-Monzote, F., Godfray, H. C. J., Griffon, M., Gypmantisiri, P., Haddad, L., Halavatau, S., Herren, H., Holderness, M., Izac, A., Jones, M., Koohafkan, P., Lal, R., Lang, T., McNeely, J., Mueller, A., Nisbett, N., Noble, A., Pingali, P., Pinto, Y., Rabbinge, R., Ravindranath, N. H., Rola, A., Roling, N., Sage, C., Settle, W., Sha, J. M., Shiming, L., Simons, T., Smith, P., Strzepeck, K., Swaine, H., Terry, E., Tomich, T. P., Toulmin, C., Trigo, E., Twomlow, S., Vis, J. K., Wilson, J. and Pilgrim,

S. (2010). The top 100 questions of importance to the future of global agriculture. *International Journal of Agricultural Sustainability* 8(4), 219–236.

Quarshie, P. T., Abdulai, A.-R. and Fraser, E. D. G. (2021). Africa's "seed" revolution and value chain constraints to early generation seeds commercialization and adoption in Ghana. *Frontiers in Sustainable Food Systems* 5, 348. https://doi.org/10.3389/fsufs.2021.665297.

Quarshie, P. T., Antwi-Agyei, P., Suh, N. N. and Fraser, E. D. G. (2022). Tackling Post-Covid-19 pandemic food crises through the adoption of improved maize seeds and technologies by smallholder farmers: the case of Ejura Sekyeredumase in Ghana. *Frontiers in Sustainable Food Systems* 6. https://doi.org/10.3389/fsufs.2022.804984.

Rai, N. D., Devy, M. S., Ganesh, T., Ganesan, R., Setty, S. R., Hiremath, A. J., Khaling, S. and Rajan, P. D. (2021). Beyond fortress conservation: The long-term integration of natural and social science research for an inclusive conservation practice in India. *Biological Conservation* 254, 108888. https://doi.org/10.1016/j.biocon.2020.108888.

Ricciardi, V., Ramankutty, N., Mehrabi, Z., Jarvis, L. and Chookolingo, B. (2018). How much of the world's food do smallholders produce? *Global Food Security* 17, 64–72. https://doi.org/10.1016/j.gfs.2018.05.002.

Rockström, J., Williams, J., Daily, G., Noble, A., Matthews, N., Gordon, L., Wetterstrand, H., DeClerck, F., Shah, M., Steduto, P., de Fraiture, C., Hatibu, N., Unver, O., Bird, J., Sibanda, L. and Smith, J. (2017). Sustainable intensification of agriculture for human prosperity and global sustainability. *Ambio* 46(1), 4–17. https://doi.org/10.1007/s13280-016-0793-6.

Rotz, S., Fraser, E. D. G. and Martin, R. C. (2019). Situating tenure, capital and finance in farmland relations: implications for stewardship and agroecological health in Ontario, Canada. *The Journal of Peasant Studies* 46(1), 142–164. https://doi.org/10.1080/03066150.2017.1351953.

Rotz, S., Gravely, E., Mosby, I., Duncan, E., Finnis, E., Horgan, M., LeBlanc, J., Martin, R., Neufeld, H. T., Nixon, A., Pant, L., Shalla, V. and Fraser, E. (2019). Automated pastures and the digital divide: how agricultural technologies are shaping labour and rural communities. *Journal of Rural Studies* 68, 112–122.

Smetana, S., Palanisamy, M., Mathys, A. and Heinz, V. (2016). Sustainability of insect use for feed and food: Life cycle assessment perspective. *Journal of Cleaner Production* 137, 741–751. https://doi.org/10.1016/j.jclepro.2016.07.148.

Smith, A., Snapp, S., Chikowo, R., Thorne, P., Bekunda, M. and Glover, J. (2017). Measuring sustainable intensification in smallholder agroecosystems: A review. *Global Food Security* 12, 127–138. https://doi.org/10.1016/j.gfs.2016.11.002.

Struik, P. C. and Kuyper, T. W. (2017). Sustainable intensification in agriculture: The richer shade of green. A review. *Agronomy for Sustainable Development* 37(5), 39. https://doi.org/10.1007/s13593-017-0445.

Suttie, D. (2020). Overview: rural poverty in developing countries – Issues, policies and challenges. *Retrievable*. Available at: https://www.un.org/development/desa/dspd/wpcontent/uploads/sites/22/2020/03/Suttie-Paper.pdf.

Taifouris, M. and Martin, M. (2021). Toward a circular economy approach for integrated intensive livestock and cropping systems. *ACS Sustainable Chemistry and Engineering* 9(40), 13471–13479. https://doi.org/10.1021/acssuschemeng.1c04014.

Taylor, C. A. and Rising, J. (2021). Tipping point dynamics in global land use. *Environmental Research Letters* 16(12), 125012. https://doi.org/10.1088/1748-9326/ac3c6d.

Thornton, P. K. (2010). Livestock production: Recent trends, future prospects. *Philosophical Transactions of the Royal Society of London. Series B, Biological Sciences* 365(1554), 2853–2867. https://doi.org/10.1098/rstb.2010.0134.

Tilman, D., Balzer, C., Hill, J. and Befort, B. L. (2011). Global food demand and the sustainable intensification of agriculture. *Proceedings of the National Academy of Sciences of the United States of America* 108(50), 20260–20264.

Tilman, D., Cassman, K. G., Matson, P. A., Naylor, R. and Polasky, S. (2002). Agricultural sustainability and intensive production practices. *Nature* 418(6898), 671–677. https://doi.org/10.1038/nature01014.

Tomlinson, I. (2013). Doubling food production to feed the 9 billion: A critical perspective on a key discourse of food security in the UK. *Journal of Rural Studies* 29, 81–90. https://doi.org/10.1016/j.jrurstud.2011.09.001.

Tscharntke, T., Clough, Y., Wanger, T. C., Jackson, L., Motzke, I., Perfecto, I., Vandermeer, J. and Whitbread, A. (2012). Global food security, biodiversity conservation and the future of agricultural intensification. *Biological Conservation* 151(1), 53–59. https://doi.org/10.1016/j.biocon.2012.01.068.

Tuminello, J. A. (2019). Environmental and animal pragmatism. In: Kaplan, D. M. (Ed.). *Encyclopedia of Food and Agricultural Ethics* (pp. 707–713). Springer Netherlands. https://doi.org/10.1007/978-94-024-1179-9_293.

Turner, K. G., Odgaard, M. V., Bøcher, P. K., Dalgaard, T. and Svenning, J.-C. (2014). Bundling ecosystem services in Denmark: Trade-offs and synergies in a cultural landscape. *Landscape and Urban Planning* 125, 89–104. https://doi.org/10.1016/j.landurbplan.2014.02.007.

United Nations Declaration on the Rights of Indigenous Peoples. Sixty-first session. Resolution adopted by the General Assembly on 13 September 2007. Available at: https://undocs.org/A/RES/61/295.

van Dijk, M., Morley, T., Rau, M. L. and Saghai, Y. (2021). A meta-analysis of projected global food demand and population at risk of hunger for the period 2010–2050. *Nature Food* 2(7), 494–501. https://doi.org/10.1038/s43016-021-00322-9.

van Huis, A. (2013). Potential of insects as food and feed in assuring food security. *Annual Review of Entomology* 58, 563-583. https://doi.org/10.1146/annurev-ento-120811-153704.

van Ittersum, M. K., van Bussel, L. G. J., Woolf, J., Grassini, P., van Wart, J., Guilpart, N., Claessens, L., de Groot, H., Weibe, K., Mason-D'Croz, D., Yang, H., Boogaard, H., van Oort, P. A. J., van Loon, M. P., Saito, K., Adimo, O., Adjei-Nsiah, S., Agali, A., Bala, A., Chikowo, R., Kaizzi, K., Kouressey, M., Makoi, J. H. J. R., Ouattara, K., Tesfaye, K. and Cassman, K. G. (2016). Can sub-Saharan Africa feed itself? *Agricultural Sciences* 113(52), 14964–14969. https://doi.org/10.1073/pnas.1610359113.

Vanlauwe, B., Coyne, D. L., Gockowski, J., Hauser, S., Huising, J. E., Masso, C., Nziguheba, G., Schut, M. and van Asten, P. J. A. (2014). Sustainable intensification and the African smallholder farmer. *Current Opinion in Environmental Sustainability* 8, 15–22. https://doi.org/10.1016/j.cosust.2014.06.001.

WEF (2019). The world is facing a $15 trillion infrastructure gap by 2040. Here's how to bridge it. *Retrievable*. Available at: https://www.weforum.org/agenda/2019/04/infrastructure-gap-heres-how-yosolveit/#:~:text=The%20world%20is%20facing%20a,bridge%20it%20%7C%20World%20Economic %20Forum.

Weiler, A. M. and Grez, E. E. (2022). Rotten asparagus and just-in-time workers: Canadian agricultural industry framing of farm labour and food security during the

COVID-19 pandemic. *Canadian Food Studies/la Revue Canadienne des Études sur l'Alimentation* 9(2), 38–52.

Weis, T. (2013). *The Ecological Hoofprint: The Global Burden of Industrial Livestock*. Wittman: Bloomsbury Publishing.

Wittman, H., James, D. and Mehrabi, Z. (2020). Advancing food sovereignty through farmerdriven digital agroecology. *International Journal of Agriculture and Natural Resources* 47(3), 235–248. https://doi.org/10.7764/ijanr.v47i3.2299.

World Bank Group (2022). Sustainable Intensification of Agriculture Project. Accessed April 1, 2022. Available at: https://projects.worldbank.org/en/projects-operations/project-detail/P007315.

WWDR (2018). *World Water Assessment Programme (Nations UNIES)*. The United Nations World Water Development Report. New York: United Nations Educational, Scientific and Cultural Organization.

Yin, X., Chen, J. and Li, J. (2022). Rural innovation system: revitalize the countryside for a sustainable development. *Journal of Rural Studies* 93, 471–478.

Chapter 10

Key ethical issues in livestock farming

Natalie Thomas, University of Guelph, Canada; and Adam B. D. Langridge, Nipissing University, Canada

1 Introduction

A presupposition of animal ethics is it has practical application in agriculture. Theoretical animal ethics uses reasoning and logic to generate moral principles and moral laws as they pertain to the relationship between humans and non-human animals and can be applied in concrete circumstances. Few farmers may be able to identify major ethical theories by name, but such theories underlie their thinking and decision-making. Zimdahl (2022) notes that the utilitarian ethical theory in particular is taken for granted in many of the formal and informal codes of ethics in agriculture; explicit awareness of utilitarianism and the wider range of ethical theories applicable to animals would increase the clarity of the principles and thinking already in place. It would make the assumptions that underlie moral problems visible and facilitate the reduction of moral harm. For animal ethicists, on the other hand, agriculture is a central concern insofar as the sector is one of the main spaces relationships between humans and animals are manifest. Moreover, climate change, food security issues, loss of farmers and food producers, the advent of biotechnologies, and genetic engineering (Fischer, 2020; Have and Neves, 2021; Marie, 2006; Meijboom et al., 2021; Thomas, 2022; Thompson, 2020a; Turner and D'Silva, 2006; Zimdahl, 2022) are only a few of the ethical issues that stem from animal agriculture. Approaches to such problems in agriculture are of general ethical importance because their effects are global. In short, those involved in animal agriculture need theoretical animal ethics to do their jobs, and animal ethicists need to pay attention to agriculture to do their jobs.

http://dx.doi.org/10.19103/AS.2023.0125.11

To assist in providing an appropriate ethical basis for studying animal ethics in agriculture, the following chapter provides a brief overview of the main ethical issues related to the use of animals. Section 1 provides a necessary, brief overview of key moral theories along with an explanation of how these inform the main positions in animal ethics, each position located on a continuum between abolitionism, the complete prohibition of the use of animals for human purposes, and what can be called the open-use position, which views non-human animals as entirely at the service of human ends. After a quick mention of the importance of religious and cultural values in animal ethics, Section 2 provides a discussion of the major issues stemming from the use of animals in agriculture into three categories: (1) issues specific to non-human animals, (2) those related to humans, and (3) those related to the effects of animal agriculture on the natural environment. Areas covered in the section on non-human animals are (i) the moral permissibility of their use in agriculture, (ii) issues stemming from industrial agriculture and factory farming, (iii) issues stemming from drugs and genetic engineering, and (iv) food security issues stemming from the combination of disease and supply-chain fragility. Areas covered in the section on humans include (i) exploitative labour practices, (ii) moral injury and trauma, and (iii) food insecurity. Finally, the section on the natural environment includes (i) pollution and greenhouse gas emissions, (ii) wilderness loss, and (iii) species loss.

2 Overview of ethical theories and positions in animal ethics

2.1 Ethical theories

Although explicit consideration for the moral status of non-human animals is not entirely absent from history even if such consideration is only obliquely implied by placing humans at the centre of it all, animal ethics as a defined subdiscipline in ethics is a recent event. In animal ethic's short existence, almost every ethical theory has been reinterpreted and applied to it, but utilitarian and deontological theories have dominated the discussion. To frame the discussion of ethical issues in animal agriculture, brief overviews of utilitarianism and deontology are therefore needed. This will be followed by a few words on other notable theoretical dispositions in animal ethics.

2.1.1 Utilitarianism

The fulcrum of utilitarianism is the principle of utility, which is basically the idea that whatever results in a net increase in happiness conceived of as pleasure or welfare is morally good and whatever results in a net decrease in happiness or welfare or an increase in suffering is morally bad or evil. A main reason given in favour of the principle of utility comes from empirical observation: every living thing is attracted to what causes pleasure and is repelled by what causes pain, meaning that there is a general intuition shared by all living things that pleasure is good and pain is bad. Utilitarianism merely makes explicit the sort of reasoning used implicitly by all sentient beings. What this means is that members of animal species apart from humans need to be considered in moral reasoning. Humans may be open to forms of pleasure and pain closed to other animals, such as the pleasure of reading about animal ethics, but it is pleasure and pain in general that matter, and all animals feel pleasure and pain.

Peter Singer (1975, 2006) developed the implicit theory of animal ethics already present in classical utilitarianism more fully. According to him, all animals should be given moral consideration insofar as they possess relevant interests as sentient beings whose welfare can be helped or harmed by human uses of them. In terms of livestock, Singer's view would result in the equal consideration of interests for all those involved (humans and animals) in practices involving animals as food, labour, and clothing, and although this would certainly reduce or curtail many practices that cause suffering to animals, it would not necessarily lead to total condemnation of the use of animals in agriculture (Singer, 1975, 2006). Harvesting an animal may be morally acceptable so long as the animal's welfare is preserved during its life and so long as the slaughter is humane.

2.1.2 Deontology

Classical or Kantian deontology, named after its main proponent, Immanuel Kant (1724-1804), is non-consequentialist, meaning that effects on welfare have no bearing on morality. What Kant calls goodwill is the starting point of the theory, and such is a will determined to do what's right for no other reason than that it is right. We are obliged to do the right thing from duty, which is the recognition that a moral imperative or a universal law of behaviour applies to us. Recognition of duty and moral law entails being able to comprehend laws and represent them in thought and language. Only humans, so far as Kant seems aware, are capable of this, which makes humans special. Humans are autonomous persons; anything a human does presupposes an intention to act in a certain way according to a principle or abstract value. We implicitly act in concert in a moral community to create moral law and abstract values. Everything else, including non-human animals, only has value, moral or otherwise, in relation to humans and the moral community. We can say, for instance, that animals have legal rights or that it is wrong to harm them, but we should not overlook that it is humans positing such rights and protections on animals. The animals themselves have no conception of legal rights, and in other epochs, there would be nothing amiss in holding animals to have no rights or moral worth at all, as history demonstrates.

Although his theory is clearly anthropocentric, it should be noted that Kant himself endorsed the kind treatment of animals when possible to develop human moral capacities; moreover, Kant's original position has been further developed to extend moral status to animals. Korsgaard (2011, 2018), for instance, has reinterpreted Kant to endorse direct moral duties to animals based on the shared 'animal nature' between humans and other animals. She argues that we should reduce and perhaps completely cease many uses of animals for human purposes due to the status of animals as ends-in-themselves (Korsgaard, 2018). In another instance, Regan (1983, 2003) argued that the relevant criterion of autonomy and personhood is not strictly rationality as Kant saw it; it is, rather, being 'subject-of-a-life'. What this means is that any entity with subjective experiences would de facto have the status of person and could therefore not be reduced to a mere instrument for human ends. We would be obligated to regard their ends as equal to our own.

2.1.3 Additional theoretical dispositions in animal ethics (ecofeminism, virtue ethics, eudaimonism, contractarianism and pragmatism)

In addition to the positions informed by utilitarianism and deontology, a few other perennial ethical theories applied to animal ethics are worthy of note. McKenna (2018), for instance, provides a 'pragmatic ecofeminist' view on living with livestock that acknowledges our deeply connected relationships with animals that she believes cannot be simply ignored. Gruen (2015) provides another example of an ethical perspective that rejects justice-based theories of utilitarianism and deontology in favour of a variety of care ethics she coins 'entangled empathy', focusing on the specific relationship contexts of ethical issues between humans and animals.

Virtue ethics informs another strain of theory in animal ethics. Such theories focus on the traits or characteristics of the person committing the action in question, where a virtuous person will intuitively commit virtuous actions and avoid vice. Virtue ethics originated with Aristotle in the ancient Greek context, but modern variations of this theory in relation to animals and nature have been expanded on by Hursthouse (2011), Diamond (2001), and Sandler (2010). The idea is that we ought to treat animals and nature with ethical consideration according to how such treatment either enriches or detracts from our own moral character. Actions promote good moral habits or degrade them, and we should want good moral habits because we should want to be good people. Virtue-based ethical theories can be useful in considering how the undue suffering of animals in agricultural practices might affect our moral integrity as humans.

Another Aristotle-inspired set of ethical theories focuses on eudaimonia or animal 'flourishing'. Nussbaum (2007, 2011) claims that a good life based on various capacities inherent to species should be the basis of our ethical consideration of animals. Rollin (2006) claims the value of each animal is located in its species' unique *telos* or function, from which ethical norms specific to different animal species can be derived. Rollin has had a notable effect on veterinary ethics and on views related to the treatment of livestock. Both Nussbaum's and Rollin's positions also consider how animals living in captivity are affected negatively in relation to their species-specific capabilities and traits. Their theories have been influential in a variety of applications and animal-related industries including zoos and animal agriculture (Gruen, 2011; Taylor, 2009; Thomas, 2022).

Contractarianism, the view that morality is the function or effect of implicit or explicit agreements provides another theoretical lens through which to view animal ethics. Contractarianism can be used to support the idea that domesticated animals benefit from living with humans within a type of contract that provides a mutual reward (Milligan, 2015), which would mean that animals are to be included in moral consideration. And, in John Rawls' version of the theory, the idea that social contracts can be formed to maximize justice from behind a veil of ignorance is seen by some to be inclusive of animals. Others, however, view the inability of animals to recognize agreements or contracts as meaning animals are de facto incapable of morality and completely outside of moral consideration. So, while contractarianism is used to support domestication as a modified form of contract between animals and humans, the theory is also often simplified in such a way as to justify intuitively immoral practices by humans. The domestication-as-contract view does not provide much by way of ethical guidance in our treatment of animals, especially given the increasing complexity of issues we are faced with currently (Garner, 2005; Milligan, 2015; Taylor, 2009).

In addition to the general disposition towards the ethics of animal agriculture that simply extends perennial moral theories to this specific subdomain, some take a more pragmatic approach beginning with animal ethics practitioners: farmers or agricultural scientists who work directly with animals. Thompson (2015b) challenges the view that animal ethics in agriculture is simply the application of moral theory to specific empirical areas of study, implying it instead to be more of a practical or procedural form of ethics that essentially includes the participation of non-specialists. Philosophers and bioethicists, the reasoning goes, tend to focus more on theoretical debates and meta-ethical questions of value and miss relevant contextual features; even those within the field of animal ethics tend to apply a favoured theoretical approach to larger level issues that often are unable to resolve real and pressing ethical issues in concrete agricultural practices. Some, however, agree with Thompson that animal ethics in agriculture should be more pragmatic but claim that ethical theory is still needed to frame on-the-ground practice. For instance, Bawden (2007) argues that access to theory for those who work in the field would promote a needed increase in the ability to be self-critical and less reductionist which could ultimately result in a reduction of moral harm. Finding a balance between the theoretical and pragmatic and/or procedural approaches to addressing ethical issues that arise in this field is important and is itself an ongoing issue in animal ethics as it applies to agriculture and beyond.

2.2 The continuum of moral consideration for animals

Given the various ethical theories and perspectives presented above, it should be clear that there will be differing implications for morally acceptable ways for humans to treat animals. One helpful way of organizing the various positions is to consider each as being roughly placeable on a continuum organized according to the extent to which animals can be used by humans (Thomas and Langridge, 2021). What follows will briefly outline this continuum of views.

2.2.1 Anthropocentrism and open use

Anthropocentric views, some of which derive from religious views, and some from ethical theories such as Kantian deontology, consider non-human animals to have no moral value in themselves, meaning that they can be used in whatever way humans see fit. This is one extreme end of the spectrum, anthropocentrism, or what can be called the open-use position, simply meaning that any use we wish to put animals to that serves human ends is ultimately morally open.

2.2.2 Animal welfarism

Based on utilitarianism, animal welfarism generally considers the use of animals for human purposes as acceptable so long as there is adequate consideration of wellbeing (Webster, 2005a, b). However, how this plays out in determining exactly what is permissible to which type of animal in what contexts is a bit more complicated. We are morally obligated to take the welfare of animals into account in whatever we do, and some who adopt an animal welfare perspective see this as meaning that all sentient creatures, be they humans or squirrels, are owed equal moral consideration (Singer, 1975, 2006). Others, however, note species of animals are open to different sets of pleasures and pains, and, therefore, each species has different moral interests. A considerable amount of scientific work has been needed to identify precisely measurable markers that represent forms or ways in which animals feel pain and suffer. Such knowledge is needed to increase welfare through things like habitat enrichment or the discovery of less painful animal harvesting procedures (Dawkins, 2012).

Moreover, even in instances where animal suffering is clear, it must be weighed against human interests, such as those held by farmers, consumers, and even future generations (Sorensen et al., 2001), and such weighing ought to avoid the horns of begging-the-question in favour of human interests and avoiding real human suffering caused by scarcity of food and clothing that come from animals (Norwood and Lusk, 2011; Sandoe et al., 2006), and it should take into account the need to adequately address ethical issues related to the ongoing use of animals, especially in agriculture (Schmidt, 2011). Difficulties in deciding acceptable use or weighing human and animal welfare are exacerbated by the problem that those who work directly with agricultural animals, including veterinarians, animal scientists, farmers, and the consumers of animal products, possess very different viewpoints on animals and their welfare, such that there may in practice be no unified view of animal welfare (Mench, 2020).

So long as an animal is given a good life according to its species, many see harvesting animals for food and clothing as permissible from a utilitarian perspective; others who give more weight to the pain of death, see the use of animals in any way as a form of suffering, and thus lean towards the prohibition of the use of animals for any reason. And so, a range of possible uses of non-human animals by humans spreads out from the animal welfare perspective even when there is agreement about the sorts of moral considerations owed to the animals involved (Garner, 2005).

2.2.3 Abolitionism

On the opposite end of the spectrum from anthropocentrism and the open-use position is the abolitionist position, which sees any use of non-human animals for purely human ends as wrong. As mentioned in the previous paragraph, there is a utilitarian basis for such a position. Perhaps ironically, however, modified versions of Kantian deontology are typically used to argue animal rights positions. Kant saw the ability to reason as the key determinant of autonomy and personhood; others have posited different criteria to expand personhood to almost all sentient creatures. Some claim that it is really the lived experience of a sentient creature that is what we implicitly see as valuable. What makes murdering a human wrong, for instance, is that we take a life, which is not merely a life in the biological sense, but is an entire lived experience. To be consistent with that intuition, any animal of a species capable of being the subject-of-a-life should be considered an autonomous person and a de facto member of the moral community, with a set of moral rights equal to the set possessed by humans (Regan, 1983, 2003). Exactly which animal species have subjective lived experiences may be debatable; however, it is clear that most species of animals used by humans in agriculture do, which would mean we are morally prohibited from continuing such use.

Another strategy for achieving similar ends is to minimize the cognitive requirements for autonomy. We can never use humans as mere ends according to Kant because humans are autonomous or capable of self-directed action. Kant believes only the ability to represent and adhere to universal law, which is tantamount to cognitive capacity to reason, provides autonomy, but even without the faculty of reason strictly speaking there are many animal species that express preferences through their actions, meaning that they are also capable of self-directed action and of valuing, and are therefore autonomous, self-aware agents (Korsgaard, 2018; Thomas, 2016). Again, exactly which animal species are capable of expressing preferences through behaviour may be open to debate, but it is clear that animals used for food and clothing are in that group. Such animals would be members of the moral community and subject to morality of the same type as humans, in some cases implying a complete prohibition of their use for almost any purpose other than companionship (Francione, 2008; Regan, 1983, 2003), and in other cases allowing for mutually beneficial relationships that protect the autonomy of humans and other animals (Korsgaard, 2018; Thomas, 2016).

2.3 The importance of religious and cultural values relevant to animals

In addition to the spectrum of positions between open use and abolition, animal ethics involves awareness of cultural and religious values related to animals and interspecies relationships. In North America, for instance, the family farm with the red barn remains the paradigm of agriculture to most, and a set of animals and their use is presupposed to go along with this. Chickens are used for eggs and meat, cows for meat and clothing, horses for labour and recreation, pigs for meat, dogs for labour and security, etc. Raising dogs for food and clothing in the North American context would cause moral outrage. To be clear, such values are culturally relative; dogs are for food in some other cultures, and the common notion in North America that a dog could be a family member would seem absurd in other contexts.

Religion also poses values on non-human animals and the natural world more generally. Some religions view pigs as in some ways polluted and unfit for human consumption; others consider eating pigs acceptable. Of religious views on animal ethics, the doctrine of dominion from the Book of Genesis (1:28) is worth singling out for attention, because some consider it uniquely responsible for the beliefs that have informed the use of non-human animals and the natural world in North American and European cultures. God proclaimed the first humans to have dominion over all of creation. Some dominionists take this to mean that it is permissible to use non-human animals in whatever way we see fit. They were created for the purpose of fulfilling our needs. Rather than see non-human animals as completely disposable to humane ends, other dominionists see the doctrine as implying a stewardship role over creation (De Mello, 2012). Parents have dominion over their children, but they would be misusing the role and the responsibility that goes along with it if they forced their children into backbreaking labour or use them as slaves to produce food. Rather, dominion (as stewardship) means that we have a responsibility to care for and maintain creation, which means helping other animals achieve their natural ends. Indeed, some religious beliefs imply that because animals are at our mercy, we are obligated to avoid causing them suffering and should adopt vegetarian or vegan diets (Linzey, 2009).

To sum up so far, animal ethics in both theory and practice has been dominated by utilitarian and deontological positions, although almost every major theory in ethics has been applied to the human-animal relationship. In terms of actual use, it is helpful to consider a continuum of available positions with the extremes of anthropocentrism which entails the open use of animals by humans and abolitionism which entails the complete prohibition of the use of animals for purely human ends. Moreover, animal ethics in any applied context cannot proceed without an understanding of the cultural and religious values that inform peoples' intuitions about the human-animal relationship. This brief and regrettably incomplete picture of animal ethics is required to move to the next section, which is an overview of the main ethical issues stemming from the use of animals in agriculture.

3 Overview of issues in animals in agriculture

The following issues are organized into groupings, but it is important to note that such groupings are imperfect and there is overlap between groups; several issues could rightly be placed in more than one group. Moreover, most issues are not discreet - they are clusters of sub-issues. For the most part, the items listed below outline a basic conundrum that gives rise to moral controversies in context. The most obvious group of issues deals with non-human animals, and so those are treated first, followed by issues related to the humans that work with livestock in various capacities, and the effects of animal agriculture on the natural environment.

3.1 Issues related to animals

3.1.1 The use of animals

The first set of issues has to do with the very presence of animals in agriculture. Many take their use as livestock and labour for granted and see such use is ingrained in their cultural way of life and in the economy. However, such a view was also taken in the not-too-distant past with historical wrongs such as slavery. If non-human animals do have intrinsic moral worth, then it is perhaps impermissible to use them as mere means to human ends in agriculture. Their use for labour could be tantamount to slavery and the reduction of some species to livestock used for food and clothing is exploitative in the least but could be considered theft and murder.

Another way of framing this issue is to consider the contradictory conception of livestock animals as both commodities and sentient beings (Wilkie, 2014). Though views on animal domestication may see benefits to both the animals and the humans involved in agriculture, there are also concerns that considering animals as property reduces domesticated animals to nothing more than products, especially for consumers who are distanced from the actual interactions with these animals themselves. Within the context of agricultural economics, there is a steady pressure to increase efficiency and productivity that can result in the increasing reduction of animals to commodities, which reduces the ability of farmers to make ethical choices they may desire in relation to their treatment of the animals they tend to (McMullen, 2016). As our knowledge of the mental and emotional capacities of animals has grown, however, more concern has been raised by ethicists and animal protection groups that livestock are not just things, but are individuals worthy of living their lives free from confinement and exploitation for profit. The philosophical and ethical issue then surrounds not just the moral status of these animals, but also their metaphysical nature. Are they commodities, or are they sentient beings deserving of rights in some form? These and similar issues are increasingly being analysed in a number of fields, including animal or human-animal studies, business ethics, and animal ethics (Ellis, 2014; Fulfer and Clipsham, 2022; Thomas, 2022).

Others within the field of animal ethics take a more moderate approach than abolitionists and claim that we should primarily be concerned with animal welfare. Humans developed as omnivores, and disagreement exists on whether humans can maintain physical health without consuming some meat (Armelagos, 2010; Pollan, 2007). Moreover, entire sectors of the economy depend on the use of animals and the products they provide. It would be possible to minimize using animals, but the economic hardship caused by simply ending current agricultural practices altogether would mean devastating and intolerable suffering for humans. Moreover, there is the possibility that many livestock species would go extinct were their use prohibited. The livestock species found on the modern farm are a product of the relationship with humans and never have existed in the wild, and such relationships can be respectful, mutually enriching, and beneficial (McKenna, 2018). Humans may therefore bear the responsibility to ensure the survival of livestock species; outside of human farming practices it is implausible to see such species persisting but for a few examples in zoos and kept as pets. For these reasons, many animal welfare advocates claim it is permissible to use animals in agriculture if a good life for the species and individuals is ensured, and if the animal is killed in as painless and humane manner as possible, then the killing of the animal and the use of it for food and clothing is morally acceptable. It maximizes utility.

Others argue that individualist ethical theories - utilitarian and deontological simply don't apply to animals. We should focus on the 'domestic contract' between animals and humans to ground a moral theory of animal welfare (Larrere and Larrere, 2000). Finally, it is worth noting as Thompson (2015b, 2020b) points out, there is often a reluctance in animal welfare discussions to engage in issues related to animal welfare other than to condemn animal agriculture outright, either in its industrial form or otherwise, rather than to insist on the need for reforms or modifications that would improve the welfare of livestock animals. Given the increasing demand for animal food products on a global scale, it is imperative that ethicists be engaged to assist in finding appropriate responses to issues as they inevitably arise in agriculture (Gremmen, 2017). Once again, this is a reminder of the mutual need for greater cross-discipline communication and collaboration between animal advocates and agricultural science, which supports animal welfare but often lacks 'a vocabulary with which to discuss ethical norms' (Thompson, 2017, p. 31) and animal ethics, broadly speaking.

On the furthest an opposite end of the spectrum from animal abolitionists, we will find those who either assume anthropocentrism, or who by omission, reduce animals to mere objects. By doing so, animals are viewed and treated merely as a means to maximizing the production capacities of animals for the economic benefits that also maximize human welfare.This can be considered an open-use position. It is in response to this position, perceived or real, that many critiques of animal agriculture are directed (Hodges, 2003).

3.1.2 Industrial animal agriculture and factory farming

Related to this debate is the distinction between so-called industrial or factory farming and other forms of agricultural practice. Concentrated Animal Feeding Operations, or CAFOs, house billions of animals in the United States alone, and these tend to be the focus of many ethical critiques due to their perceived effect on the welfare of the animals (Halteman, 2011; MacClellan, 2014; Rossi and Garner, 2014). This is despite responses like those from Thompson (2015b) who says that it's not a necessary conclusion from the conditions within CAFOs that those animals would be better off dead. For him, and others, this also implies that we should just give up and all become vegetarians or vegans as it's implied by the nature of the critiques that improvements simply can't be made in such institutions. The implementation of high-density feed lots, industrial chicken coops, and technology like rotolactors, for instance, presupposes that the primary value of animals is in the form of commodities only, and not as creatures with emotions or sensations that can be harmed. This view is increasingly being adopted on a global scale, intensifying concerns about the implications on agricultural systems and the animals and humans within them (Shiva, 2006). Those who assume the open-use position see only the benefits for humanity in industrial forms of agriculture, but both abolitionists and some animal welfare advocates see the use of animals in factory farming as inherently immoral. Other forms of agriculture that explicitly consider the welfare of animals, such as small-scale and organic farms, are seen to maximize utility and are therefore morally praiseworthy (Bruckner, 2020). Even abolitionists consider such types of agriculture the lesser of two evils and advocate them for the sake of practicality. It is unlikely that any argument will win consumers and businesses that depend on products from animals or on the use of animals over to abolitionism despite the existence of non-animal alternatives. Humane forms of animal agriculture are at least improvement from factory farming that would still allow business and consumer demand for the products of animal agriculture to be satisfied.

If animals can be considered stakeholders with humans who care about their welfare and those who seek to profit from them, it is a potentially novel way to frame ethical considerations for animals (Lerner et al., 2013; Smart, 2022). Given the influence and importance of consumer perceptions on animal welfare in relation to their consumption of and preferences for different animal food products, this should also be considered when devising codes of ethics and statements of values by agricultural companies and corporations that rely on the use of animals (Bell et al., 2017; Bunting and Galyean, 2015; Frawley and Dyson, 2014; McKendree et al., 2014). The explicit inclusion of animal welfare incorporated in mission statements, vision statements, and ethical codes can also be considered an important aspect of developing corporate social responsibility more fully and transparently among animal agricultural industries which would strengthen the ethical reputation of such companies and their practices to consumers and the rest of society (Janssens and Kaptein, 2016; Maloni and Brown, 2006).

3.1.3 Livestock and biotechnologies

The use of hormones, antibiotics, and gene editing to enhance the productivity of livestock and to shorten the production cycle increases efficiency and productivity, which in turn has economic benefits in the form of more abundant and affordable food and clothing. It is therefore seen by many as a moral good that promotes human welfare. However, hormone, antibiotic use, and gene editing come with their own sets of ethical issues. Not only do such practices sometimes not take the welfare of non-human animals into account, but they have potentially harmful effects on humans (Anomaly, 2020). The hormones and drugs given to enhance animal growth are consumed by humans directly or indirectly through the waste that seeps into waterways used for drinking water. Gene-editing technologies (CRISPR) have made the manipulation of species' genetic codes to produce animals better capable of serving human ends both affordable and accessible, but perhaps not acceptable. Gene editing may not cause immediate physical dangers to humans through consumption of genetically modified animals - in fact, meat from genetically modified food is safe. However, some see gene-modified animals as a threat to species integrity - a Frankenfood, or dystopian. Consumers choose to avoid such foods as frightening even if not physically harmful.

The moral implications of gene editing extend beyond the potential safety issues. What makes the use of animals in agriculture and by humans more generally morally wrong is that such use disregards or outright violates the moral dignity of sentient beings (De Graeff et al., 2019; Streiffer and Basl, 2011; Shriver, 2020). High-density chicken coops, for instance, cause pain and suffering to the chickens involved and violate the animals' autonomy (Thomas, 2016). However, in the not-too-distant future gene-editing techniques may advance to the point that sentience or aspects of it, such as the ability to feel pleasure and pain or cognitive abilities sufficient for autonomy, could be edited out of a species' genetic code. This would make the animals more easily manageable and would thus have an economic benefit. It would also ignore/sidestep moral principles used to argue against the use of animals in agriculture (Shriver, 2009; Shriver and McConnachie, 2018). It is impossible to violate autonomy and cause pain if the species in question lacks sentience, pain receptors, or basic awareness (Thompson, 2008). Many see this as wrong and dystopian. Such disenhancement might still violate autonomy; if there is a positive obligation to promote autonomy, and the principle of humanity seems to say there is, then disenhancement, which precludes autonomy and sort of cuts it off at the stem, would be morally impermissible (Thomas and Langridge, 2022).

Gene editing, the use of hormones and other chemical manipulations, and recent technological innovations may change the way we think about animals. Currently, there is a trend to reduce animals used in agriculture primarily to livestock and as commodities, such as when species are bred for increased muscle mass at the expense of other aspects of their health and wellbeing. In the near future, however, there may be a further move to view the biological aspects of agriculture (food production, clothing, or any other use) as technology rather than interspecific relationships. This could take the ethical debate out of its current context. To explain, meat fit for human consumption can be grown in laboratories and the source of such meat would not be animals. It is difficult to harm an animal in meat production if the meat production does not use animals, to begin with. And it is worth keeping in mind that the stated purpose of the project to create lab-grown meat is to provide an ethical source of meat to that portion of vegans and vegetarians who are because of their moral objections to factory farming and the perceived mistreatment of animals. Moreover, lab-grown meat does not necessarily threaten the existence or integrity of species currently used as food. Small-scale or family farm production that uses traditional non-modified breeds of cattle, poultry, and other species could continue alongside scaled-up lab-grown meat production. Even if lab-grown meat entirely replaced beef, poultry, and pork from traditional species, the existence of such species could still be maintained. Pigs are now kept as pets, and horses, once kept primarily for transportation and for use as draught animals, live in the wild and are still used by humans for equestrian sports and other recreational uses in addition to companionship. So, perhaps in a few years, we will be talking about ethical issues in biotech rather than animals in agriculture, and such a shift will bring with it a completely different set of ethical considerations.

3.2 Issues related to humans

3.2.1 Labour

It should be noted that the primary focus on the use of non-human animals by humans in the area of ethical issues involving animals in agriculture is warranted, because the relationship between humans and other species has been for the benefit of humans and to the detriment of other species, especially in agriculture. However, it should not be overlooked that humans are animals and that they are involved with other animals in agriculture. In fact, it might be more useful to consider agriculture as an impersonal system with implicit values that inform behaviours; both human and non-human species of animals are reduced to roles within that system. What humans supply to the system is labour, and in some contexts, they can be reduced to this capacity in the same way that animals can be reduced to commodities (Porcher, 2011). In this way, ethical labour practices could be considered an inherent aspect of animal ethics in agriculture.

3.2.2 Moral injury and trauma

Although social justice issues inherent in the labour practices of agriculture are a genuine set of ethical issues related to animals in agriculture, the more germane position of humans in this topic is via interspecies personal relationships. Non-human animals have clearly defined purposes in agriculture that lie outside of companionship, but growing cattle involves caring for them, and caretakers develop genuine and sometimes deep connections with the animals they raise and work with. Some of the most meaningful relationships people can have are with the pets they care for, and, even if steps are taken to remain cold and detached, in the daily process of looking to the needs of livestock relationships just as meaningful can form. Knowing that ultimately these animals will be slaughtered or that painful procedures must be performed on them can result in the compartmentalization of moral feelings and emotional desensitization (Humphries, 2022).

Damage to humans can be in their work with non-human animals and when an emotional attachment to the animals is lacking. Workers tasked with harvesting chickens from high-density coops and then cleaning the coops encounter deformed chickens and sometimes those half-eaten by other chickens. It is a mix of excreta, blood, and flesh. Working in such conditions can be horrifying, and some develop post-traumatic stress and other psychological disorders (Grusnovik and Blaznik, 2022). Given our increasing understanding of the emotional and psychological implications of working under such conditions with animals, the promotion of the wellbeing of livestock aids human wellbeing.

In addition to the psychological and emotional harms humans can incur from working with livestock, there is also the potential to incur damage to each individual's moral character. The level of comfort in harming non-human animals that can develop in those who work with animals in agriculture transfers to comfort in violence towards humans, and this concept is present in both virtue ethics and Kantian ethical theory. The use of violence and physical force to control and harm non-human animals gradually turns to a general disposition towards using violence and physical force to solve problems. In addition to the suffering caused to animals and potentially to humans through physical and psychological harm, the character of the individuals involved thus suffers. Any society should want to be a good society, and a good society is one that promotes excellence of character in all of its members. Not only is the viciousness that develops in the individuals immediately involved in violence towards animals in agriculture wrong in itself, but the system that promotes viciousness over virtue is therefore morally wrong.

3.2.3 Food security

In addition to the issues with factory farming related to suffering and indignity it can cause to animals are issues of food security and human health, some of which have been mentioned in the previous section. The economic benefits are obvious justifications for factory farming, but purely economic justifications are based on moral ones. The use of modern manufacturing processes, technology, and intensification of production may lead to cheaper, more abundant food, undeniable social benefits, and increase in the overall welfare of humans, especially in developed countries. Factory farming may eliminate food scarcity by providing more abundant food sources and reduced prices, if it makes food more accessible to consumers across socio-economic groups. It is ironic that factory farming's justification also makes a major issue evident. High-density production lots and farms are ideal conditions for the growth of the disease, and the globalized distribution system facilitates the rapid spread of diseases. Mad cow disease is the most obvious example. Some of the effects of mad cow disease have been the destruction of entire populations of cattle in countries in which the disease occurs, accompanied by job loss and a large rise in food costs. The breakdown in global supply chains caused by mad cow disease also has negative effects, including small, local producers, being pushed out of the market because they are unable to compete with large industrial production organizations. In the event of a supply-chain breakdown, local suppliers needed to fill the gap may no longer exist. In addition to this, the 'farming poor', which, as Thompson (2015a) notes, are most often women on a global scale, need to sell what they grow to survive. This means, for Thompson, that the moral value of farming needs to be emphasized because it promotes economic agency, and this addresses not only food security and poverty, but can also be indirectly used as the justification for adopting something like a 'One Health' approach into livestock farming. The justification of industrial-scale meat production and global distribution systems thus pose threats to human welfare through their inherent fragility.

The use of costly and increasingly scarce food resources for raising livestock is another way the use of animals in agriculture affects humans. Given that the global demand for animal food products is expected to double in the year 2050, the issue of food scarcity and the rising human population of the world means that the pressure to feed all inevitably leads to the intensification of farming practices (Scholten et al., 2013). Concerns such as these have led to the development of the 'One Health' platform that links the health of animals, humans, and nature to ensure food security in less affluent countries (Thumbi et al., 2015). In many such locations, animal agriculture assists humans in achieving food and economic security, which also means that the health of the livestock becomes of paramount importance for fulfilling basic life needs. If animal agriculture is damaging to the environment, then the harms of more intensive practices can harm humans by damaging overall sustainability (Buller et al., 2018).

Another way to view this is to consider agriculture has only one central purpose: to produce food for humans. Cattle were domesticated and are still kept primarily for beef, chickens are raised for poultry and eggs, and pigs are for pork. Obviously, they need to be fed, and the resources that go into creating feed, some would argue, would best be used directly for human food. Many positions in animal ethics on livestock are in some way related to this issue, with most conclusions reflecting the moral superiority of reducing or eliminating reliance on animal sources of food, in part due to the negative effects on humans and nature through the necessary exploitation of resources to sustain such industries (Kemmerer, 2020; Singer, 2006). Invariably, livestock indirectly competes with humans for food and water, and such competition raises prices humans pay for all sorts of food as we increasingly experience the effects of climate change and more intensive strains on resources through the rise of the global human population.

3.2.4 Zoonotic diseases and pandemics

Zoonotic diseases are also an issue with factory farming. In a post-pandemic world not paying attention to diseases that develop in animals we eat or have contact with can have serious effects on our health and wellbeing. Intensive food production practices (high-density feed lots) encourage creation, spread, and mutation of diseases. As Sebo (2022) notes, the use of antimicrobials to stimulate livestock growth and prevent the spread of disease creates ideal conditions for antimicrobial-resistant mutations to pass among animals and human workers. Humans work with animals, and should a disease with the ability to infect humans develop, the opportunity to move to human hosts is present in industrial-scale operations and high-density feed lots. It is not the case that the risk posed by zoonotic disease is not present in other forms of agriculture; it is amplified in industrial animal farming.

3.3 Issues related to the natural environment

3.3.1 Moral consideration for the environment and sustainability

Some argue that while the natural environment or an ecosystem is neither sentient nor a subject-of-a-life, it still has intrinsic moral worth. We know, or we ought to know, what it means to bring harm to an ecosystem. It is morally wrong to do something to wantonly pollute a river. However, many either don't take such arguments seriously or find them irrelevant, assuming that the worth of sentient beings always outweighs an ecosystem or river. Such thoughts are always secondary to the needs sentient beings have and therefore never make a difference. However, even if we have no immediate responsibilities to the natural environment, all can agree that the environment does deserve consideration because sentient life depends on it. Polluting a river may not be wrong in itself, but it violates the autonomy and personhood of sentient entities by rendering the water unfit for the consumption of animals, including humans (Thomas, 2016). Hence, for some, any harm to the environment is wrong. For others, sustainability, or the ability to continue a practice indefinitely without catastrophic harm sufficient to lead to termination of the practice, is enough. Environmental ethics perspectives lay out different views on the ethical relationship between humans and the natural environment, and within these, we find, as we do in animal ethics, a continuum of views from purely anthropocentric that sees nature as instrumentally valuable, to Deep Ecology or the Land Ethic, which view nature as inherently valuable (Thomas and Langridge, 2021). In this section, there are inevitable crossovers with previous issues related to animal agriculture that involve animals and humans.

In terms of sustainability, there are challenges not only in defining what this means but also in determining whether or not livestock farming can be done in ways that improve environmental quality and human and animal health. For example, Broom et al. (2013) argue that silvopastoral systems can not only be sustainable but can improve animal welfare and enrich biodiversity. McKenna (2018) similarly shows, through conversations with farmers who have diverse livestock, that grazing ecosystems that work with the surrounding natural environment instead of reshaping it is not the problem. Rather, large-scale industrial or 'segmented' farming is. This does not address the larger issue of growing global demand for animal products that may make it difficult to support smaller-scale yet sustainable operations. It does however show the importance of recognizing the interdependent nature of sustainability itself that links together the health of humans, animals and the environment with science, ethics, and public policy in addressing food security issues (Buller, 2018). It shows the interconnected nature of the ethical issues throughout this chapter, as gene technology, while presenting its own ethical issues, can also potentially be one way to resolve food scarcity and hunger, and yet may have potentially detrimental effects on the environment (Thompson, 2021). The concept of sustainability is important to 'One Health' in addressing the combined ethical concerns regarding livestock farming on animals, humans, and nature, while also remaining problematic in its variable definition and ambiguous application.

3.3.2 Pollution and greenhouse gas emissions

The issues stemming from the use of animals in agriculture in relation to the environment for the most part have to do with the perceived negative effects of agricultural practices, or the environmental damage, they cause. Although agriculture has always affected the natural environment in which it takes place, scale and intensity matter. The ability of a river to neutralize the harmful effects of pollution from farm runoff through dissipation, for instance, cannot be maintained indefinitely. There is a line, however fuzzy, between older and more traditional forms of farming and agriculture and the modern industrial-scale varieties. Although the by-products of livestock production are for the most part used in agriculture, chemicals such as nitrogen and phosphorous and waste runoff make their way into groundwater, rivers, lakes, and other bodies of water (Kemmerer, 2020; Zimdahl, 2022). Pollution also results also from the increasing amount of land for farming to meet the demand for animal food products. It will require more fertilizer, pesticides, and water. Animal agriculture consumes over 50% of all available freshwater (Sebo, 2022). Animal agriculture pollutes freshwater and consumes it in large amounts, depleting the world's freshwater preserves (Kemmerer, 2020).

Another key environmental effect of animal agriculture is the emission of greenhouse gases (methane, carbon dioxide, and nitrous oxide) (Kemmerer, 2020). Moreover, deforestation to provide more land for growing feed for livestock and grazing results in a decrease in carbon storage sinks (Sebo, 2022). The Amazonian wildfires of 2019 and clearing land through 'slash and burn' techniques can quickly become out of control, harming humans, the forest, and wildlife, as well as contributing to climate change. Many argue if a reduction in greenhouse gases sufficient to stave off catastrophic global warming is to be avoided, then such practices need to be eliminated to reduce atmospheric carbon dioxide (MacClellan, 2014). They are unsustainable and unethical due to their effects on the biosphere. In addition to this, climate change will make it more challenging to maintain current agricultural practices. Increased soil erosion, disease, insect pests, and water shortages will further strain the food supply and increase the costs of food, particularly those derived from livestock.

3.3.3 Species loss and biodiversity

Clearing land for cattle production and industrial animal farms (CAFOs) and grains grown for feed creates species habitat loss. This is an issue wherever large-scale animal agriculture occurs, and it means that habitats for wildlife and plants are destroyed in a way that reduces overall biodiversity. Biodiversity is valuable to maintain ecosystem integrity which benefits all life on the planet, mitigates climate change, and maintains essential pollinating insects. Increasing global deforestation for livestock is antithetical to environmental conservation (Broom et al., 2013). In addition to the species loss from habitat loss, species considered predators of livestock animals have been hunted to, at times, near extinction, and attempts to reintroduce such species results in tension and a clash of interests between livestock producers and conservationists over the issue. This in turn can lead to an imbalance among species within an ecosystem and eventually lead to ecosystem collapse. As Hedberg (2022) notes, biodiversity is valued for its aesthetic appeal and because it maintains an ecosystem which provides products and services humans need. Given the threat that climate change also poses to biodiversity, there is a powerful ethical impetus to protect biodiversity through as many approaches as possible, with some arguing that livestock production either needs to cease completely, or be dramatically reduced, or fundamentally changed to sustainable and wildlife-friendly farming methods (Broom et al., 2013).

4 Conclusion

This chapter organized the main issues in the use of animals in agriculture into three general but overlapping categories relevant to moral consideration: (1) non-human animals, (2) humans, and (3) the natural environment. The main controversial moral issue relevant to non-human animals has to do with their moral status and the permissibility of using sentient non-human animals for human ends. There are myriad positions. The dominant ones are abolitionism, animal welfarism, and open use. The recent advent of affordable genetic manipulation technologies raises further ethical issues because it may be possible to create livestock without sentience and meat and other animal products in isolation of the whole animal. Another major ethical issue is exploitative labour practices. One must also consider the psychological and emotional trauma of some who work with animals. Moreover, the use of animals as livestock creates social justice issues related to access to food in addition to food prices stemming from meat production. In addition, the natural environment, pollution, wilderness loss, species loss, and increased greenhouse gases related to animals in agriculture have clear ethical implications.

Humans have not always considered themselves essentially distinct or separate from the animals used in agriculture for food, labour, and security. We saw ourselves as unilaterally bending animals to our needs through animal husbandry and eugenics. Scholarship has shown the relationship runs both ways. Non-human animals are an essential part of civilization, not external to it. They have shaped us as we shaped them. Considering them undeserving of moral consideration as societies develop is to misunderstand what civilization disregarded - the deep relationship with other species that stands at the centre of what it means to be human. It seems quite possible, if not likely, that the use of animals in agriculture will be changed as genetics and biotech develop. Nevertheless, we should not ignore our moral responsibility to animals used in agriculture and other sentient creatures. Precisely what that entails and how interspecies relationships will look like in the future remains to be seen.

5 Where to look for further information

A good introduction to many of the issues raised in this chapter is *Animals and business ethics by* N. Thomas (see Thomas, 2022 in the References section for full details).

One central source for locating individuals who work on and perform research in the field of animal ethics and who publish both an academic journal and book series is *The Oxford Centre for Animal Ethics* (https://www.oxfordanimalethics.com/home/).

In relation to the One Health approach, which focuses on the importance of examining the interconnectedness between animals, humans and the environment, the following organizations are a useful starting point for pursuing current research in this area:

- The World Organisation for Animal Health: https://www.woah.org/en/home/.
- The Food Ethics Council (based in the UK): https://www.foodethicscouncil.org/.
- The One Health Commission (based in the USA): https://www.onehealthcommission.org/en/why_one_health/.

Research centres provide opportunities for studying animal welfare and for access to research conferences and collaboration opportunities, including the following:

- The Campbell Centre for the Study of Animal Welfare, University of Guelph: https://www.uoguelph.ca/ccsaw/.
- The National Farm Animal Council (based in Canada, but provides resources for the US too): https://www.nfacc.ca/.

5.1 Key journals and books

- *Animal Welfare*: https://www.ufaw.org.uk/the-ufaw-journal/animal-welfare.
- *Journal of Applied Animal Welfare Science*: https://www.tandfonline.com/journals/haaw20.*Animals*: https://www.mdpi.com/journal/animals.
- *Journal of Animal Ethics*: https://www.press.uillinois.edu/journals/?id=jane.
- *Journal of Agricultural and Environmental Ethics*: https://www.springer.com/journal/10806.*Animal Frontiers*: https://academic.oup.com/af.

- Fischer, B. (2020). *The Routledge Handbook of Animal Ethics* (B. Robert and W. Fischer, Eds.). Routledge.
- Thomas, N. (Ed.). (2022). *Animals and Business Ethics*. Basingstoke: Palgrave Macmillan.
- Thompson, P. B. (2015). *From Field to Fork: Food Ethics for Everyone*. Oxford University Press. https://doi.org/10.1093/acprof:oso/9780199391684.001.0001.
- Thompson, P. B. (2020). *Food and Agricultural Biotechnology in Ethical Perspective* (3rd ed.). Springer. https://doi.org/10.1007/978-3-030-61214-6.
- Thompson, P. B. and Kaplan, D. M. (2014). *Encyclopedia of Food and Agricultural Ethics* (1st ed., Vol. 2014). Springer Netherlands. https://doi.org/10.1007/978-94-007-0929-4.
- Zimdahl, R. L. (2022). *Agriculture's Ethical Horizon* (3rd ed.). Elsevier.

6 References

Anomaly, J. (2020). Intensive animal agriculture and human health. In: Fisher, B. (Ed.). *The Routledge Handbook of Animal Ethics*. Routledge, pp. 167–176. https://doi.org/10.4324/9781315105840-16.

Armelagos, G. J. (2010). The omnivore's dilemma: The evolution of the brain and the determinants of food choice. *Journal of Anthropological Research* 66(2), 161–186. https://doi.org/10.3998/jar.0521004.0066.201.

Bawden, R. (2007). Redesigning animal agriculture: A systemic perspective. In: Swain, D. L., Charmley, E., Steel, J. and Coffey, S. (Eds.). *Redesigning Animal Agriculture*. CABI Publishing, pp. 1–17. https://doi.org/10.1079/9781845932237.0001.

Bell, E., Norwood, F. B. and Lusk, J. L. (2017). Are consumers wilfully ignorant about animal welfare? *Animal Welfare* 26(4), 399–402.

Broom, D. M., Galindo, F. A. and Murgueitio, E. (2013). Sustainable, efficient livestock production with high biodiversity and good welfare for animals. *Proceedings: Biological Sciences* 280(1771), 20132025. https://doi.org/10.1098/rspb.2013.2025.

Bruckner, D. W. (2020). Small-scale animal agriculture. In: Fisher, B. (Ed.). *The Routledge Handbook of Animal Ethics*. Routledge, pp. 198–210. https://doi.org/10.4324/9781315105840-18.

Buller, H., Blokhuis, H., Jensen, P. and Keeling, L. (2018). Towards farm animal welfare and sustainability. *Animals: An Open Access Journal From MDPI* 8(6). https://doi.org/10.3390/ani8060081.

Bunting, L. D. and Galyean, M. L. (2015). Customer and consumer confidence in the livestock industry–Professional ethics. *Professional Animal Scientist* 31(4), 309–314. http://dx.doi.org/10.15232/pas.2015-01399 also see - Bunting, L., & Galyean, M. L. (2015). INVITED PAPER: Customer and consumer confidence in the livestock industry–Professional ethics. *The Professional Animal Scientist* 31(4), 309–314. https://doi.org/10.15232/pas.2015-01399.

Dawkins, M. S. (2012). *Why Animals Matter: Animal Consciousness, Animal Welfare, and Human Well-Being*. Toronto: Oxford University Press.

De Graeff, N., Jongsma, K. R., Johnston, J., Hartley, S. and Bredenoord, A. L. (2019). The ethics of genome editing in non-human animals: A systematic review of reasons reported in the academic literature. *Philosophical Transactions of the Royal Society of London. Series B, Biological Sciences* 374(1772), 20180106. https://doi.org/10.1098/rstb.2018.0106.

De Mello, M. (2012). *Animals and Society: An Introduction to Human-Animal Studies*. New York: Columbia University Press.

Diamond, C. (2001). *The Realistic Spirit*. Cambridge, MA: MIT Press.

Ellis, C. (2014). Boundary labor and the production of emotionless commodities: The case of beef production. *The Sociological Quarterly* 55(1), 92–118. https://doi.org/10.1111/tsq.12047.

Fischer, B. (2020). *The Routledge Handbook of Animal Ethics*. New York: Routledge.

Francione, G. L. (2008). *Animals as Persons: Essays on the Abolition of Animal Exploitation*. New York: Columbia University Press.

Frawley, J. K. and Dyson, L. E. (2014). Animal personas: Acknowledging non-human stakeholders in designing for sustainable food systems. In: *Proceedings of the 26th Australian Computer-Human Interaction Conference on Designing Futures: The*

Future of Design (OzCHI '14). New York: Association for Computing Machinery, pp. 21–30. https://doi.org/10.1145/2686612.2686617.

Fulfer, K. and Clipsham, P. (2022). Are animals always commodified in the context of business? In: Thomas, N. (Ed.). *Animals and Business Ethics*. Basingstoke: Palgrave Macmillan, pp. 21–41.

Gremmen, H. G. J. (2017). *Ethics of Livestock Farming?: Who Cares?* Wageningen University & Research. https://doi.org/10.18174/417430.

Gruen, L. (2011). *Ethics and Animals: An Introduction*. New York: Cambridge University Press.

Gruen, L. (2015). *Entangled Empathy: An Alternative Ethic for Our Relationships With Animals*. Brooklyn: Lantern Books Book Company.

Halteman, M. C. (2011). Varieties of harm to animals in industrial farming. *Journal of Animal Ethics* 1(2), 122–131. https://doi.org/10.5406/janimalethics.1.2.0122.

Have, H. T. and Neves, M. C. P. (2021). Agricultural ethics. In: *Dictionary of Global Bioethics*. Springer International Publishing, pp. 105–106. https://doi.org/10.1007/978-3-030-54161-3_49.

Garner, R. (2005). *Animal Ethics*. Malden: Polity Press.

Grusnovik, T. and Blaznik, M. (2022). Denied relationship: Moral stress in the vocational killing of non-human animals. In: Thomas, N. (Ed.). *Animals and Business Ethics*. Basingstoke: Palgrave Macmillan, pp. 251–270.

Hedberg, T. (2022). Animal suffering, environmental impact, and lab-cultured meat. In: Thomas, N. (Ed.). *Animals and Business Ethics*. Basingstoke: Palgrave Macmillan, pp. 179–202.

Hodges, J. (2003). Livestock, ethics, and quality of life. *Journal of Animal Science* 81(11), 2887–2894. https://doi.org/10.2527/2003.81112887x.

Humphries, R. (2022). Moral feelings, compartmentalization and desensitization in the practice of animal experimentation. In: Thomas, N. (Ed.). *Animals and Business Ethics*. Basingstoke: Palgrave Macmillan, pp. 229–250.

Hursthouse, R. (2011). Virtue ethics and the treatment of animals. In: Beauchamp, T. L. and Frey, R. G. (Eds.). *The Oxford Handbook of Animal Ethics*. Toronto: Oxford University Press, pp. 119–143. https://doi.org/10.1093/oxfordhb/9780195371963.013.0005.

Janssens, M. R. E. and Kaptein, M. (2016). The ethical responsibility of companies towards animals: A study of the expressed commitment of the Fortune Global 200. *Journal of Corporate Citizenship* 2016(63), 42–72. http://doi.org/10.9774/GLEAF.4700.2016.se.00005.

Kemmerer, L. (2020). Animal agriculture and the environment. In: Fisher, B. (Ed.). *The Routledge Handbook of Animal Ethics*. Routledge, pp. 154–166. https://doi.org/10.4324/9781315105840-15.

Korsgaard, C. M. (2011). Interacting with animals: A Kantian account. In: Beauchamp, T. L. and Frey, R. G. (Eds.). *The Oxford Handbook of Animal Ethics*. Toronto: Oxford University Press, pp. 91–111. https://doi.org/10.1093/oxfordhb/9780195371963.013.0004.

Korsgaard, C. M. (2018). *Fellow Creatures: Our Obligations to the Other Animals*. New York: Oxford University Press.

Larrere, C. and Larrere, R. (2000). Animal rearing as a contract? *Journal of Agricultural and Environmental Ethics* 12(1), 51–58. https://doi.org/10.1023/A:1009552109479.

Lerner, H., Algers, B., Gunnarson, S. and Nordgren, A. (2013). Stakeholders on meat production, meat consumption and mitigation of climate change: Sweden as a case.

Journal of Agricultural and Environmental Ethics 26(3), 663–678. https://doi.org/10.1007/s10806-012-9420-0.

Linzey, A. (2009). *Why Animal Suffering Matters: Philosophy, Theology, and Practical Ethics*. New York: Oxford University Press.

MacClellan, J. (2014). Animal agriculture and welfare footprints. In: Thompson, P. B. and Kaplan, D. M. (Eds.). *Encyclopedia of Food and Agricultural Ethics*. Springer Netherlands. https://doi.org/10.1007/978-94-007-0929-4.

Maloni, M. J. and Brown, M. E. (2006). Corporate social responsibility in the supply chain: An application in the food industry. *Journal of Business Ethics* 68(1), 35–52. https://doi.org/10.1007/s10551-006-9038-0.

Marie, M. (2006). Ethics: The new challenge for animal agriculture. *Livestock Science* 103(3), 203–207. https://doi.org/10.1016/j.livsci.2006.05.006.

McKendree, M. G. S., Croney, C. C. and Widmar, N. J. (2014). Effects of demographic factors and information sources on United States consumer perceptions of animal welfare. *Journal of Animal Science* 92(7), 3161–3173.

McKenna, E. (2018). *Livestock: Food, Fiber, and Friends*. Athens, GA: University of Georgia Press.

McMullen, S. (2016). *Animals and the Economy*. Basingstoke: Palgrave Macmillan.

Meijboom, F., Yalim, N. Y. and Evren, M. (2021). Animals in transition: On ethics and livestock farming. In: *Proceedings of the Book of 3rd International Congress on Agricultural and Food Ethics*, pp. 85–90.

Mench, J. A. (2020). Animal welfare–Is intensification the problem? In: Fisher, B. (Ed.). *The Routledge Handbook of Animal Ethics*. Routledge, pp. 141–153. https://doi.org/10.4324/9781315105840-14.

Milligan, T. (2015). *Animal Ethics the Basics*. New York: Routledge.

Norwood, F. B. and Lusk, J. L. (2011). *Compassion, by the Pound: The Economics of Farm Animal Welfare*. New York: Oxford University Press. https://doi.org/10.1093/acprof:osobl/9780199551163.001.0001.

Nussbaum, M. C. (2007). *Frontiers of Justice; Disability, Nationality, Species Membership*. Cambridge: The Belknap Press.

Nussbaum, M. C. (2011). The capabilities approach and animal entitlements. In: Beauchamp, T. L. and Frey, R. G. (Eds.). *The Oxford Handbook of Animal Ethics*. Toronto: Oxford University Press, pp. 228–252. https://doi.org/10.1093/oxfordhb/9780195371963.013.0009.

Pollan, M. (2007). *The Omnivore's Dilemma: A Natural History of Four Meals*. Penguin Books.

Porcher, J. (2011). The relationship between workers and animals in the pork industry: A shared suffering. *Journal of Agricultural and Environmental Ethics* 24(1), 3–17. https://doi.org/10.1007/s10806-010-9232-z.

Regan, T. (1983). *The Case for Animal Rights*. Berkeley: University of California Press.

Regan, T. (2003). *Animal Rights, Human Wrongs: An Introduction to Moral Philosophy*. Toronto: Rowman & Littlefield Publishing Group.

Rollin, B. E. (2006). *Animal Rights & Human Morality* (3rd ed.). New York: Prometheus Books.

Rossi, J. and Garner, S. A. (2014). Industrial farm animal production: A comprehensive moral critique. *Journal of Agricultural and Environmental Ethics* 27(3), 479–522. https://doi.org/10.1007/s10806-014-9497-8.

Sandler, R. (2010). Ethical theory and the problem of inconsequentialism: Why environmental ethicists should be virtue-oriented ethicists. *Journal of Agricultural and Environmental Ethics* 23(1-2), 167–183.

Sandoe, P., Christiansen, S. B. and Forkman, B. (2006). Animal welfare: What is the role of science? In: Turner, J. and D'Silva, J. (Eds.). *Animals, Ethics, and Trade: The Challenge of Animal Sentience*. Routledge, pp. 41–52.

Schmidt, K. (2011). Concepts of animal welfare in relation to positions in animal ethics. *Acta Biotheoretica* 59(2), 153–171. https://doi.org/10.1007/s10441-011-9128-y.

Scholten, M. C. T., De Boer, I. J. M., Gremmen, B. and Lokhorst, C. (2013). Livestock farming with care: Towards sustainable production of animal-source food. *NJAS* 66(1), 3–5. https://doi.org/10.1016/j.njas.2013.05.009.

Sebo, J. (2022). *Saving Animals, Saving Ourselves: Why Animals Matter for Pandemics, Climate Change, and Other Catastrophes*. Toronto: Oxford University Press.

Shiva, V. (2006). The implications of agricultural globalization in India. In: Turner, J. and D'Silva, J. (Eds.). *Animals, Ethics, and Trade: The Challenge of Animal Sentience*. Routledge, pp. 193–207.

Shriver, A. (2009). Knocking out pain in livestock: Can technology succeed where morality has stalled? *Neuroethics* 2(3), 115–124. https://doi.org/10.1007/s12152-009-9048-6.

Shriver, A. (2020). Genetic engineering of nonhuman animals. In: Fisher, B. (Ed.). *The Routledge Handbook of Animal Ethics* (1st ed.). Routledge, pp. 267–272. https://doi.org/10.4324/9781315105840-24.

Shriver, A. and McConnachie, E. (2018). Genetically modifying livestock for improved welfare: A path forward. *Journal of Agricultural and Environmental Ethics* 31(2), 161–180. https://doi.org/10.1007/s10806-018-9719-6.

Singer, P. (1975/2009). *Animal Liberation*. New York: HarperCollins.

Singer, P. (Ed.). (2006). *In Defense of Animals: The Second Wave*. Malden: Blackwell Publishers.

Smart, J. (2022). Animals as stakeholders. In: Thomas, N. (Ed.). *Animals and Business Ethics*. Palgrave Macmillan, pp. 297–324.

Sorensen, J. T., Sandoe, P. and Halberg, N. (2001). Animal welfare as one among several values to be considered at farm level: The idea of an ethical account for livestock farming. *Acta Agriculturae Scandinavica* 30 (Suppl 30), 11–16. https://doi.org/10.1080/090647001316922992.

Streiffer, R. and Basl, J. (2011). Ethical issues in the application of biotechnology to animals in agriculture. In: Beauchamp, T. L. and Frey, R. G. (Eds.). *The Oxford Handbook of Animal Ethics*. Toronto: Oxford University Press, pp. 826–854. https://doi.org/10.1093/oxfordhb/9780195371963.013.0031.

Taylor, A. (2009). *Animals & Ethics: An Overview of the Philosophical Debate* (3rd ed.). Peterborough: Broadview Press.

Thomas, N. (2016). *Animal Ethics and the Autonomous Animal Self*. Basingstoke: Palgrave Macmillan.

Thomas, N. (Ed.). (2022). *Animals and Business Ethics*. Basingstoke: Palgrave Macmillan.

Thomas, N. E. and Langridge, A. B. D. (2021). Animal ethics and business ethics. In: Poff, D. C. and Michalos, A. C. (Eds.). *Encyclopedia of Business and Professional Ethics*. Cham: Springer. https://doi.org/10.1007/978-3-319-23514-1_1255-1.

Thompson, P. B. (2008). The opposite of human enhancement: Nanotechnology and the blind chicken problem. *NanoEthics* 2(3), 305–316. https://doi.org/10.1007/s11569-008-0052-9.

Thompson, P. B. (2015a). *From Field to Fork: Food Ethics for Everyone*. Toronto: Oxford University Press. https://doi.org/10.1093/acprof:oso/9780199391684.001.0001.

Thompson, P. B. (2015b). Livestock welfare and the ethics of producing meat. In: Thompson, P. B. (Ed.). *From Field to Fork*. Toronto: Oxford University Press, pp. 130–158. https://doi.org/10.1093/acprof:oso/9780199391684.003.0006.

Thompson, P. B. (2017). The ethics of food animal production. In: Kalof, L. (Ed.). *The Oxford Handbook of Animal Studies*. Oxford: Oxford University Press, pp. 363–379. https://doi.org/10.1093/oxfordhb/9780199927142.013.26.

Thompson, P. B. (2020a). *Food and Agricultural Biotechnology in Ethical Perspective* (3rd ed.). Springer. https://doi.org/10.1007/978-3-030-61214-6.

Thompson, P. B. (2020b). Philosophical ethics and the improvement of farmed animal lives. *Animal Frontiers: the Review Magazine of Animal Agriculture* 10(1), 21–28. https://doi.org/10.1093/af/vfz054.

Thompson, P. B. (2021). Food system transformation and the role of gene technology: An ethical analysis. *Ethics and International Affairs* 35(1), 35–49. https://doi.org/10.1017/S0892679421000034.

Thompson, P. B. and Kaplan, D. M. (Eds.). (2014). *Encyclopedia of Food and Agricultural Ethics*. Springer Netherlands. https://doi.org/10.1007/978-94-007-0929-4.

Thumbi, S. M., Njenga, M. K., Marsh, T. L., Noh, S., Otiang, E., Munyua, P., Ochieng, L., Ogola, E., Yoder, J., Audi, A., Montgomery, J. M., Bigogo, G., Breiman, R. F., Palmer, G. H. and McElwain, T. F. (2015). Linking human health and livestock health: A 'one-health' platform for integrated analysis of human health, livestock health, and economic welfare in livestock dependent communities. *PLoS ONE* 10(3), e0120761.

Turner, J. and D'Silva, J. (2006). *Animals, Ethics, and Trade: The Challenge of Animal Sentience*. Routledge.

Webster, J. (2005a). *Animal Welfare: Limping Towards Eden: A Practical Approach to Redressing the Problem of Our Dominion Over the Animals*. Blackwell Publishers.

Webster, J. (2005b). Ideals and realities: What do we owe to farm animals? In: Turner, J. and D'Silva, J. (Eds.). *Animals, Ethics and Trade: The Challenge of Animal Sentience*. London: Earthscan Publications, pp. 149–158.

Wilkie, R. (2014). Animals as sentient commodities. In: Kalof, L. (Ed.). *The Oxford Handbook of Animal Studies*. Toronto: Oxford University Press, pp. 279–301. https://doi.org/10.1093/oxfordhb/9780199927142.013.16.

Zimdahl, R. L. (2022). *Agriculture's Ethical Horizon* (3rd ed.). Elsevier.

Chapter 11

Ethical issues of developing new technologies in agriculture

Bart Gremmen, Wageningen University, The Netherlands

1 Introduction

Contemporary society faces a series of grand challenges on topics like transport, energy production, and healthcare. Innovation is the preferred solution for resolving these grand challenges, and, therefore, the concept has become 'the emblem of the trendy society' (Godin, 2015). An important grand challenge is the anticipated rise of the worldwide demand for agricultural products (FAO, 2012). However, agriculture already faces a range of problems, regarding the reduced welfare of animals, the utilization of pesticides, susceptibility to infectious diseases, housing systems, soil degradation, and the high use of antibiotics and vaccines. These problems will decrease the number of obtainable future agricultural products. Could most of those problems be solved by a series of innovations?

Incremental innovation has always been at the center of agriculture (Klerkx and Rose, 2020). From the start of agriculture, farmers have developed ways to use plants and animals for human consumption. Step by step, using trial-and-error breeding processes, they slowly domesticated crop plants and animals. This process of thousands of years of domestication has led to an expansion of cultivated fields, with only a few crop plants that are produced in monocultures, and to the use of mainly five species of animals in livestock farming (ibid). The

http://dx.doi.org/10.19103/AS.2023.0125.12

results of incremental innovation are integrated quite easily into existing agro-ecosystems. Any ethical issues are immediately dealt with. In case of any bad outcomes, it is still possible to reverse the changes or interventions.

In the last centuries, incremental innovation has continued and agricultural research has led to several radical innovations in agriculture. These innovations, starting from mechanization to hybrid cultivars and to the development of modern biotechnology, have shaped conventional agriculture on a world scale (Zambon et al., 2019). However, non-governmental organizations and policymakers have raised serious ethical issues which are the results of these radical innovations (Gremmen et al., 2019). The results of radical innovation transform existing agro-ecosystems into new ones. If any ethical issues are not immediately dealt with, the changes are often irreversible. As a consequence, agriculture has lost an enormous part of its license-to-produce from society. To regain this license, the Responsible Research and Innovation (RRI) initiative urges all radical innovators to identify and help solve moral problems in the early stages of innovation (Von Schomberg, 2013). The idea is to deal with potential ethical issues that threaten radical agricultural innovation as soon as possible in the developmental process (Grunwald, 2010). This will lead to radical, responsible reorganizations of farmers' practices.

From the perspective of agricultural ethics, I will investigate potential ethical issues threatening radical innovation in agriculture, and what kind of questions these ethical issues entail for innovators. First, the role of ethics in the development of radical new technologies in general will be described. My conclusion is that innovators have to be able to deal with emergent ethical issues in the development of new technologies. Second, the RRI initiative will be described and analyzed to apply it to agriculture. Third, four potential ethical issues will be identified for developing new technologies in agriculture. These ethical issues signal the possible occurrence of criticism from stakeholders, consumers, and policymakers and thus lead to specific normative questions innovators must answer. Two samples of developing agricultural technologies that aim to change organisms: genetic modification and the toolbox of new plant breeding technologies (NPBTs) will be analyzed.

2 Ethics and developing new technologies

In a number of innovation areas, such as genomics and synthetic biology, moral problems occur in the early stages of innovation (e.g. Singer, 1986). According to Collingridge, social control of technology is necessary to avoid the harmful social consequences of a new technology (Collingridge, 1980). This may be done by changing technology in its infancy or by imposing controls and restrictions. Two conditions are necessary to avoid the undesired consequences of a new technology: it must be known that a technology has, or will have, harmful effects, and it must be possible to change the technology in some way to avoid the effects (Collingridge, 1980). One or both of the conditions are often lacking, and attempts to control technology seldom succeed. According to Collingridge in such cases, there is a 'dilemma of control'. The first horn of the dilemma is that the harmful social consequences of the fully developed technology cannot be predicted with sufficient confidence to justify the imposition of control. The second horn of the dilemma is that by the time a technology is sufficiently well developed and dispersed its unwanted social consequences may not be apparent, it is no longer easily controlled. Control may still be possible, but it will become very difficult, expensive, and slow. What happens is that society and the rest of its technology gradually adapt to the new technology, so that, when it is fully developed, any major change requires changes in many other technologies and social and economic institutions, making its control very disruptive and expensive (Collingridge, 1980).

An important assumption of the Collingridge dilemma is the consequentialist/utilitarian perspective in ethics. The normative starting point of the dilemma is the need to avoid the harmful social consequences of a technology, but the message of the dilemma is that a consequentialist/utilitarian perspective is impossible. In the early phases of a new technology, ethical deliberations become speculative because we lack the required knowledge (Grunwald, 2010). In the latest phases of a new technology, ethical deliberations often come too late, namely, when all of the relevant decisions have already been made, when it is too late to avoid harmful consequences of the technology. Collingridge's own normative response to the dilemma is to maintain the 'freedom to control technology', because the essence of controlling technology is to retain 'the ability to change a technology, even when it is fully developed and diffused, so that any unwanted social consequences it may prove to have can be eliminated or meliorated' (Collingridge, 1980 20/21). He suggests developing organizational structures and scientific tools to deal with the resistance to such control (ibid. 19). Experts, decision makers, and end-users all are entangled in controlling the new technology.

However, Collingridge did not foresee that some experts were going to use a version of his control dilemma as a normative tool in their attempts to exclude prospective users from the innovation process. Experts sometimes stress that they are willing to include users in the early stages of the new technology (Gremmen, 2007), when there is still a lot of room to take the voice of prospective users into account in the design of the product, but the experts can offer little concrete information that would allow prospective users to imagine how they could integrate the end-product in their everyday life. This version of the Collingridge dilemma depicts the end-users in the emergence of new technologies as the end-point of a linear process. However, the world of the users and the world of technological innovation are by no means separate entities that only merge when a final product is delivered to the users; they are entangled from the start. Technology assessment, and, later, constructive technology assessment, recognized the importance of involving users in the innovation process to encourage the integration of new technologies in users' everyday lives (Rip et al., 1995; Oudshoorn and Pinch, 2003). The case has been made that technologists need to study responses to science in order to learn from them (Levitt, 2003) and to discover missing propositions in their own reasoning (Locke, 2002). Everyday life concerns that inform people's responses to emergent technologies may be at odds with scientific and technological standards but can and should be understood on their own terms. In this way, experts could benefit from the active involvement of prospective users (Veen et al., 2010).

It is difficult for innovators to deal with possible ethical issues. Not only do the results of an innovation trajectory have unknown consequences, but, more importantly, we do not know the results of innovation at the start of the innovation trajectory (Wathes et al., 2008). This means that, in moral reasoning about innovations in the making, the relevant moral facts and the appropriate principles are more or less still unknown, as are relevant moral consequences. Only by doing ethics in life sciences will the moral dilemmas in the trajectory of responsible innovation emerge.

3 Responsible research and innovation in agriculture

One of the strategies for addressing the criticism of society is stakeholder inclusion (Blok, 2019): academia, industry, citizens, and therefore the government collaborate in the innovation process to form robust, systemic solutions. Through stakeholder participation, the innovation process is predicted to become more inclusive and, as such, its results are expected to be more social-ethically accountable (Meissner and Carayannis, 2017). This 'democratization' of research and innovation is additionally called Responsible Research and Innovation (RRI) (Blok and Lemmens, 2015). For example, RRI is an emerging field within the European research and innovation (R&I) policy context that aims to balance economic, socio-cultural, and environmental aspects in innovation processes (European Commission, 2011). Because technological innovations can contribute significantly to answering societal challenges like temperature change or food security but can have negative societal consequences, it's assumed that social and ethical aspects should be considered during the R&I process. For this reason, the emerging concept of RRI concerns ethical reflection on the character, scope, and applicability of responsibility and innovation in innovation practices generally, and, therefore, the way social-ethical issues are often applied and addressed in agriculture. The concept of RRI expresses the ambition to deal with societal needs, next to the more traditional objectives like profit maximization, economic process, and competitive advantage (cf. Blok and Lemmens, 2015).

In the context of agriculture and food, RRI offers opportunities because it provides a new impulse to strengthen animal, agricultural, and food ethics. An example is the reflection on the killing of day-old male chicks (Bruijnis et al., 2015; Gremmen et al., 2018) (see Section 5). Compared to a 'classical' innovation trajectory, responsible innovation aims at some reasonable stakeholder involvement at an early stage (Blok and Lemmens, 2015; Blok, 2019). During this stage, stakeholders discuss normative values. An example of the method in agriculture is the reflexive interactive design method (Bos et al., 2009).

From a more general ethical point of view, one could argue that RRI is somewhat ambiguous. On the one hand, it promises a transparent ethical agenda (Schomberg et al., 2019; Van de Poel and Robaey, 2017)); on the other hand, it contains the danger that ethics (responsibility) is reduced to procedural categories like deliberation and responsiveness (Stilgoe et al., 2013). With living organisms, this reduction is counterproductive (Gremmen et al., 2019). To do justice to the particular assignment of RRI, ethics should, in these contexts, belong to the core of RRI (ibid). In the following section, we will elaborate on the ethically unique status of farming to enable the mainstream approach to RRI to be applied to agriculture and food. A list of four ethical issues will strengthen the moral agenda of RRI to enable the contextualization of RRI within agriculture.

4 Four ethical issues in developing new technologies in agriculture

Over the last few hundred years, there has been a plethora of innovations in farming techniques. From an ethical perspective, this is clearly very positive, but agriculture, compared to other technical practices, is special (Gremmen et al., 2019). Agriculture can be defined as the technical transformation of living material into plant and animal products, in which life processes play an essential role (ibid). Agricultural products enter our bodies, which is why farming is an interface, providing a sharp distinction between the (outside) world of nature and the (inside) world of humans. In this way, we are included in larger systems and part of many cycles. The products of farming are organic, while in all other technical practices, the products are inorganic. Without a doubt, agriculture is the most crucial human activity on the earth. Farmers don't just take care of plants and animals. They enable our survival on earth (ibid).

Without agriculture, we would be forced to collect the limited food that is available in nature. As a result, the number of people on the earth would decrease drastically. Agriculture is necessary for the survival of humanity. We can do without most other technical practices. In my view, this may explain the ethical debates about agriculture. Although we can survive without some agricultural products (e.g. meat), our culture is entirely focused on plant and meat consumption. In this way, the plants and animals 'take care' of us. Agriculture is the basis of our culture whereby, by intervening in the natural life of plants and animals, we gradually have domesticated nature. An important characteristic of this so-called incremental innovation is that it is reversible. We cannot just develop new technologies and immediately abandon agricultural processes and products in the way we can irreversibly abandon other technological processes and products. The *first ethical issue* is therefore that the development of radical technologies in agriculture is seen as disruptive by many stakeholders, citizens, and societal organizations. This means that the first ethical question for developers of new technology must be: how can we improve the current situation without irreversibly altering it?

In agricultural innovations, intrinsic normative standards seem to be provided and used, contrary to other innovations which provide no intrinsic normative standards at the start of the innovative process and call for extrinsic normative standards (e.g. in value-sensitive design). There is a symbiosis between people and domestic plants and animals in farming, as a result of which we can regard farming as a kind of 'second' nature. Animals in nature are, to a certain extent, autonomous: they do not need us to survive. We have made domestic plants and animals dependent on us by intervening in their life processes, especially their reproduction. The classic role of farmers is to take care of living things on the farm. They reproduced life by making a selection from everything that lives. Recent radical technologies in agriculture have led to the *second ethical issue*: farmers, aided by new high-tech processes, bypass the natural world and turn from reproducers of life into producers of life. The second ethical question for developers of new technology must be: are we going to change farmers into producers of life?

However, innovations leading to irreversible actions, such as the extinction of certain natural species by genetic modification, or the destruction of all the seeds of a cultivar, are very risky because interrelated cycles that continue to make the future of farming possible could be interrupted. This means that the artifacts resulting from non-agricultural innovations may have certain one-dimensional risks, while in agriculture often multi-dimensional risks may occur because of the interrelatedness of species in (agricultural) ecosystems. The need to deal with these multi-dimensional risks is the *third ethical issue*. For example, the invention of genetic modification in 1973 has been important for agriculture because it allows direct intervention in the genetic basis of the reproduction of plants and animals. This technology is transformational because it not only enables manipulation of life at the most fundamental, molecular level but also fundamentally changes the conditions of the possibility of transformation of life itself. The trial-and-error methods of the farmers are replaced by the targeted approach of scientists, therefore giving the concept of 'domestication' a completely new meaning. This means in general that the third ethical question for developers of new technology must be: can we deal with the multi-dimensional risks technologies generate?

In farming, plants and animals are not only a production goal but also a means of production. An important difference between the innovation of artifacts and animals and plants is that until now we have not designed the products in farming ourselves through a blueprint on the drawing board. There is talk of 'designing' via domestication and targeted breeding, but that is highly dependent on the biodiversity present and the natural path dependencies of the genetic material. But at this moment plants and animals in farming are not designed by us and they function relatively autonomously as means of production. In farming, the products produce themselves. They have, however, been transformed from natural organisms into domesticated organisms. This leads to the *fourth ethical issue*. Contrary to artifacts, organisms and agro-ecosystems have a double ethical status: they are both subjects and objects. In farming, there is no absolute distinction between the production process and the product. Through the process of growth, the products produce themselves, in contrast to practices in which people or machines make the products. For example, even in meat from the laboratory, the muscle cells produce the beefsteak by repetitiously dividing themselves. In farming, people only provide the preconditions and have to deal with an extra ethical status. This means that the fourth ethical question for developers of new technology must be: how must we deal with the double ethical status of the organisms?

In the following section, we'll use the four ethical issues, and their accompanying normative questions, to analyze the development of new genetic modification technologies. We'll focus on the case of the ethics of genetically modified chickens. Also, we'll use the ethical issues to find out if the technological toolbox of NPBTs, compared to genetic modification, has any specific ethical issues.

5 Ethics of genetic modification in agriculture

After the first half of the twentieth century, molecular biology developed ways to alter DNA in controlled ways. Genetic modification (GM) allowed the cutting and splicing of DNA molecules (Gremmen, 2017). Two methods were developed: the particle gun method and the agrobacteria method. These were used first in mono-cellular organisms and subsequently applied to multi-cellular organisms. The first GM-mammal, made in the mid-1970s, was a 'transgenic' mouse (a mouse containing DNA from another species). In 2015, the GM-salmon was the first GM-product to be approved for consumption in America (ibid).

The basic genetics in plants and animals is more or less the same, but the application of some methods and technologies differs (Barnes and Dupre, 2008).

For example, mutation breeding – increasing the mutation frequency through chemicals or radiation – is a common and legal method for plants but is not possible in animals, for ethical and economic reasons (Shu et al., 2011). In plant breeding, marker-assisted selection and GM have been used (Gremmen, 2005) for many years, but GM has not yet been applied to livestock on a large scale (Gremmen, 2009). The only approval given for a GM animal for food production is Canada's AquAdvantage salmon eggs, with a focus on growth enhancement (Goubau, 2011). Other examples of GM animals are pigs (i.e. increased growth rates and higher utilization of phosphate in their feed), dairy cows (enhanced resistance to mastitis, improved udder health, and improved milk quality), goats and sheep (improved wool production and disease resistance), and chickens (resistance to diseases and feed efficiency) (Thompson, 2007).

There are three different legal regimes for GM applications in the European Union (Gremmen, 2017). The industrial application of GM is the first. Industry transfers molds, viruses, and bacteria into genetically modified organisms (GMOs). This contained use is allowed because it is done in sealed containers. Agriculture is the second. The application of GM to plants is allowed. However, the application of GM to animals is not allowed, unless there are serious arguments to do it anyway. Currently, there are no GM animals allowed in the EU. The application of GM to humans is the third regime. This is the severest legal regime: GM of humans is not allowed.

Compared to the millions of hectares of soy, corn, canola, and cotton in South and North America, GM plants are almost non-existent in Europe. Although it is legal to develop and cultivate GM plants, the long and expansive regulatory road in the last decades has led to one crop in one country: about 60 000 hectares of Bt corn in Spain (ibid). The potential benefits of GMOs are higher quality, higher environmental value, lower price, and higher nutritional value. Potential societal values are higher food security, more affordable food, better food safety, higher societal health, more sustainability, and biodiversity. An important problem is that these benefits and values may clash. For example, higher food security or better price could result in lower health or lower food safety. It is all about setting the right priorities.

The results of the Eurobarometer showed in 2010 that people in the EU do not see the benefits of GM-food (Gaskell et al., 2011). Some people think GM is probably unsafe or harmful (ibid). People in the EU have a strong concern for safety and do not see the benefits of horizontal gene transfer. They have reservations about safety and the potential effects on the environment but accept the potential benefits of vertical gene transfer (ibid) (see the *third* ethical issue). Other people have strong reservations about other ethical issues, such as the use of human embryos, but think the science of regenerative medicine should be developed. Although strict laws are needed to alleviate concerns about ethical issues, they approve of stem cell research, transgenic animal research, and human gene therapy (ibid).

The ethical arguments about GMOs vary among societies (Gregorowius et al., 2012). On the critical side, some simply object to GM. It is claimed that GM technology amounts to a form of hubris concerning man's relationship to nature (are we allowed to 'play God'?) (Comstock, 2010) (see the *first* ethical issue). Criticism has also focused on other disruptive effects of GM, for example, the autonomy of farmers in deciding whether to use the technology (e.g. are patents allowed?) and the autonomy of consumers in deciding whether to put the products on the table (Comstock, 2010). An important argument against the use of GM is that it is unnatural and therefore morally problematic (Siipi, 2004; Haperen et al., 2012) (see the *second* ethical issue).

Many critics are not opposed to GM technology but to its different applications (Rollin, 2006). This means, from a consequentialist stance, that even people who do not have an objection in principle to the technology can still be critical of its use in agriculture in general, and in food production in particular (Sandoe and Christiansen, 2008). The risks and uncertainties of this new technology (see the third ethical issue) are also criticized. It is argued either that there are risks to the environment or human health, or that there might be such risks, and that for this reason some version of the precautionary principle should be applied (Gremmen, 2006).

From the viewpoint of justice, current applications of agricultural biotechnology have also been criticized, in particular with respect to the distribution of economic benefits from its use (Thompson, 2007). Also arguments about threatening the integrity and/or the intrinsic value of plants can be found (see the *fourth* ethical issue). Ethics may clarify and test such arguments and explain normative and epistemic assumptions. In livestock farming, GM may contribute to efficiency but, at the same time, may be used to circumvent ethical problems (Hanssen and Gremmen, 2013). The case about the ethics of genetically modified chickens illustrates (Bruijnis et al., 2015).

In response to the increasing demand for safe, cheap food in sufficient quantities, the intensification and mechanization of poultry farming began in the mid-twentieth century (Gremmen et al., 2018). The number of chickens kept by one farmer has increased considerably since then. Efficiency and specialization were enabled by developments in feeding, breeding, and housing of the animals and by increased knowledge of veterinary medicine, especially the availability of antibiotics for therapeutic and preventive reasons. Genetic selection enabled egg production by layer chickens and meat production by specialized meat chickens. Male chicks became less attractive for meat production (ibid). Available sexing techniques made it possible to distinguish males from females immediately after hatching, and it became common practice to kill these male day-old chicks.

In the Netherlands, over 50 million male chickens annually are killed in a gas chamber immediately after hatching. Societal opposition to this practice has prompted the development of innovations. Several alternatives to the killing of day-old chicks have been proposed (Leenstra et al., 2010); this leads to the question of whether these alternatives are morally superior. We have developed a framework to evaluate the technical and socio-ethical aspects of alternative more responsible innovations to solve this problem, selected on social desirability and technical potential compared to the current situation (Bruijnis et al., 2015). One alternative uses GM of breeding hens that permits the hatching eggs containing males to be identified with spectroscopy. This is a non-invasive technique compared to the technique of taking a sample from the egg to find the difference between male and female eggs. The GM alternative takes advantage of the genetics of birds to ensure GM-free laying hens, and also that their eggs are GM free.

This clear case of a morally inferior practice has potentially morally better alternatives. Besides the GM alternative, there are several others: raising the male chicks, dual use of chickens, taking a sample from the egg, etc. Each alternative has its advantages and disadvantages with respect to technical and socio-ethical aspects, and each has specific importance for various stakeholders. Solving one issue raised by the current situation raises new issues. For example, by acknowledging arguments against the killing of such young animals and starting to rear the males, issues arose about the effects on the environment and the marketing of the chicks. The issue of killing day-old chicks and its alternatives thus seems to be an example of choosing the least of several possible evils and can be explained by a special type of moral lock-in.

Technological lock-in has become since the mid-1980s an important subject of growing academic enquiry in the field of innovation studies, especially by economists working within an evolutionary tradition (David, 1985; Arthur, 1989). The general idea of lock-in is that technologies and technological systems follow specific paths that are difficult and costly to escape (Perkins, 2003). Even if potentially superior alternatives are available, these technologies and technological systems often survive for a very long time. An example is the triumph of the QWERTY keyboard layout over the Dvorak Simplified Keyboard layout (David, 1985). Another is the race between VHS and Betamax as a video cassette recorder standard (Arthur, 1989). Lock-in is explained by the increasing returns of an initial lead in the competition between technologies (David, 1985; Arthur, 1989). 'This arises because early adoption can generate a snowballing effect whereby the preferred technology benefits from greater improvement than its competitors, stimulating further adoption, improvement, and eventual leadership' (Perkins, 2003, 23).

There are many ways in which locked-in technologies may be inferior to their alternatives. We focus on moral lock-in: the way a production system can be locked-in to technology standards that are potentially morally inferior. In some cases, there is a consensus on the potential for moral improvement that could be achieved through the development of alternative technologies. The question then becomes: what is holding back the development of these morally better technologies? Many debates about the transition to these new technologies focus only on the costs involved (Carrillo-Hermosilla, 2013). Our hypothesis is a moral lock-in may explain the survival of morally inferior technologies. We suggest responsible innovation can be a concept for balancing economic, socio-cultural, and environmental aspects in innovation processes (EC, 2011), as an approach to morally 'unlock' alternative innovations. By involving stakeholders in the innovation process and by considering ethical and societal aspects during this process, the socio-ethical acceptability and the societal desirability of innovative products will increase significantly (Von Schomberg, 2013; Blok and Lemmens, 2015).

6 Ethics of the toolbox of new plant breeding technologies

The Court of Justice of the European Union (CJEU) decided on 25 July 2018 that all products made by the toolbox of NPBTs are GMOs. NPBTs are technologies to increase and accelerate the development of new traits in plant breeding, and gene editing, CRISPR-Cas9, is its main technology (Zhang et al., 2014). Two research institutes of the European Commission (Cf. JRC/IPTS, 2011) have described the content of the NPBTs toolbox. It contains intragenesis (technologies using transformation with genetic material restricted to the species' own gene pool), cisgenesis, emerging techniques to induce controlled mutagenesis or insertion (ODM, zinc finger nuclease technologies 1–3), and other applications such as or reverse breeding or grafting on GM-rootstocks.

In a short time, CRISPR/Cas9 has been used in plant breeding (Zhang et al., 2014). Compared to older techniques, these so-called genome editing techniques are cheaper, faster, more accurate, and more widely applicable because of their ability to cut and alter the DNA of any species at almost any genomic site with ease and precision (Jasanoff et al., 2015). They have been developed to overcome the problem of randomness that results from mutation breeding and the ability to determine the site of mutation or insertion of genes. Applications of animal gene editing techniques are many, varied, and rapidly evolving, including applications that promise benefits in welfare, disease resistance, and feed efficiency. Although gene editing technology promises significant benefits, this promise will not be realized unless the technology is firmly and fully embedded in ethical principles accepted by society. My aim is to adapt and broaden the existing ethical frameworks on biotechnology (Holland and Johnson, 1998; Rollin, 1995) to these new scientific methods and technologies. This will help scientists, stakeholders, and policymakers to understand, evaluate, and monitor the integration of the technical, social, and ethical aspects of the modern GM toolbox.

There has been a long debate in the European Union about the regulation and legal categorization of NPBTs. At the policy level, discussions have evolved around the question of whether products from NPBTs are or should be subject to special regulation (Sprink et al., 2016). Some people believe that they should also not be subject to special regulation (New Techniques Working Group, 2008) because most NPBTs could not be separated from conventional breeding techniques. Others, highlighting the requirements of the precautionary principle, call for regulation following the regulations for GMOs (Then, Bauer-Panskus, 2017). Most NPBTs are subject to regular GM regulation in the EU according to the CJEU judgment (CJEU 2018: Case C-528/16) on the mutagenesis exemption in Directive 2001/18/EC (European Parliament/Council of the European Union 2001 (hereafter Directive)). This ruling has created a regulatory system for NPBTs, which is unique in the world. If organisms obtained from NPBTs are put in the same basket as GMOs, this may carry a serious risk: transferring analogous ethical problems that GMOs encountered in the past, to organisms obtained from NPBTs, while they may not address similar risks (Poortvliet et al., 2019).

In plant breeding, several new genetic engineering techniques, also referred to as genome editing, have been developed. 'Genome editing' is 'the practice of making targeted interventions at the molecular level of DNA or RNA function, deliberately to alter the structural or functional characteristics of biological entities' (Nuffield Council on Bioethics, 2016, 4). Because of their ability to cut and alter the DNA of any species at almost any genomic site with ease and precision, these genome editing techniques are faster, more accurate, cheaper, and more widely applicable than older techniques (Jasanoff et al., 2015). They have been developed to determine the site of mutation or insertion of the genes and to overcome the problem of randomness that results from mutation breeding.

In plant breeding, the development of CRISPR-Cas9 enabled precisely targeted alterations to DNA sequences in living cells (Zhang et al., 2014). It is based on the 'virus library' of bacteria (a natural way of bacteria to defend against phage infection) and uses RNA to locate the exact spot in a genome. It is possible to insert a new piece of DNA (in the case of a cis- or transgene plant), and it cuts the unwanted piece of DNA (i.e. point mutations) (Jasanoff et al., 2015).

Applications of plant gene editing techniques are varied, many, and rapidly evolving, including applications that promise benefits in salt and drought tolerance, and disease resistance (Brandt and Barrangou, 2019). A new tomato variety that grows like a bush is one of the first cultivars. To realize the promised benefits of gene editing technology, the technology needs to be firmly and fully embedded in society. For example, plant geneticists from Tübingen have developed the 'Tomelo', a variety of tomato that is resistant to powdery mildew because it has a deletion in the SlMlo1 gene (Nekrasov et al., 2017). In less than 10 months, CRISPR-Cas9 enabled them to achieve this (ibid). The new tomato variety is also indistinguishable from naturally occurring deletion mutants and contains no foreign DNA (no natural species barrier was crossed) (Nekrasov et al., 2017).

Although there are all kinds of material on the Internet about the ethical discussion of GM, there are almost no journal papers about ethics and CRISPR-Cas9. The use of an adequate name in the societal debate is very important: is it gene editing or genetic manipulation CRISPR-Cas9 or GM? (Boersma and Gremmen, 2018). The public will link the name of a new technology to an element in the name if they lack knowledge of that new technology. For example, genomics has been linked to GM (ibid). Also, the aims and functions of the new technology are already described in different ways: tinkering with the genome; manipulation of DNA; repairing the genome; tools to create mutations; text processing of DNA. An inappropriate wording proves to be very hard to correct (Boersma et al., 2019).

How can we use CRISPR Cas9 in the genomes of plants, animals, and humans? First, the technology can be used to *repair* the genome (i.e. heritable diseases), but in some cases, a natural alternative to repair a genome also is possible. *Prevention* is a second use (i.e. inheritable diseases), and *improving* the genome (existing traits or new traits) is a third. Improvement by adding new traits offers endless possibilities. The fourth way to use this technology is to *design* new genomes. In the case of humans, this could lead to the return of earlier ethical debates about eugenics.

What are the ethical issues of CRISPR-Cas9? Because it is possible to insert (a) gene(s) through gene editing, all the ethical issues of GM apply. There are two other ethical issues specific to CRISPR-Cas9, and I conclude by linking these main technological characteristics of CRISPR-Cas9 to the first and third ethical issues in developing new technologies in agriculture:

1. The *first* ethical issue is about the possible disruptive nature of a technology. CRISPR-Cas9 is not by definition a transgene technique (Nuffield Council on Bioethics, 2016). In the media, a denial of its transgenic possibilities is used to make it more likable. Also, a point mutation deletion caused by CRISPR-Cas9 (ibid) is impossible to detect. As a consequence, the difference between GM and non-GM becomes un-detectable, thereby blocking one of the cornerstones in the regulation of genetic modification. Therefore, it will be difficult to exercise societal regulation. Because CRISPR-Cas9 is relatively fast

(Nuffield Council on Bioethics, 2016), regulation could be slow and sometimes even implemented too late. It also means that it is difficult to label products developed by this technology. New transparent and responsible chains have to be developed to ensure the consumer's right to complete information.

2. The *third* ethical issue is about multi-dimensional risks. CRISPR-Cas9 is very accurate compared to GM (Nuffield Council on Bioethics, 2016). However, side effects, like off-target mutations and unexpected results, have also been reported after certain gene editing. Also, CRISPR-Cas9 is cheap compared to GM (ibid). This makes abuse by experts and companies more possible and economically attractive. CRISPR-Cas9 is relatively easy (ibid). This makes abuse by amateurs/terrorists more conceivable. How safe is CRISPR-Cas9?

7 Conclusion

Agriculture has been dominated by incremental innovation. Only recently radical innovation has become the standard way of introducing new technologies. We started this chapter by describing the role of ethics in the development of radical new technologies in general and concluded that innovators have to be able to deal with emergent ethical issues in the development of new technologies. What does this mean for applying the RRI initiative to agriculture? We concluded that RRI contains the danger that ethics (responsibility) is reduced to a procedural category similar to deliberation and responsiveness. Our view is that, in contexts with living organisms, like agriculture, this reduction is counterproductive. To do justice to the particular assignment of RRI, ethics should be fundamental to the core of RRI. We therefore elaborated on the ethically unique status of farming to enable the mainstream approach to RRI to be applied to agriculture and food. As a result, a list of four general, potential ethical issues will strengthen the moral agenda of RRI and enable contextualization of RRI within the domain of agriculture. These ethical issues signal the possibility of criticism from stakeholders, consumers, and policymakers and thus lead to specific normative questions innovators must answer. Finally, we analyzed two samples of developing new agricultural technologies that aim to change organisms: we found all four ethical issues in the debates about GM. Also, the debates about the toolbox of NPBTs contained, the issues we already found in GM and two other ethical issues. Innovators still have plenty of opportunities to deal with these issues because the development of NPBTs is still going on. We sincerely hope that innovators will try to discover in the early stage of the development of new agricultural technologies if the ethical issues apply. And if they do, deal with the emergent normative questions as soon as possible.

8 Where to look for further information

Some organizations involved with respect to ethical issues of developing new technologies in agriculture are as follows:

1 The European Union (EU) has a special website devoted to RRI tools: https://rri-tools.eu, and the RRI-Practice project is an example of a 3-year project under Horizon 2020. Its aim is to understand the barriers and drivers to the successful implementation of RRI in both European and global contexts, to promote reflection on organizational structures and cultures of research-conducting and research-funding organizations, and to identify and support best practices to facilitate the uptake of RRI in organizations and research programs. The project will review RRI-related work in 22 research-conducting and research-funding organizations and will develop RRI Outlooks outlining RRI objectives, targets, and indicators for each organization.

The RRI-Practice project seeks to

(a) harvest experiences on how research-conducting and research-funding/policy organizations work to strengthen RRI-related values;
(b) support the systematic development of such work in these organizations;
(c) generate scalable knowledge about good practices for the wider implementation of RRI.

2 The Food and Agriculture Organization (FAO) of the United Nations provides information on ethical issues in agriculture on its website: https://www.fao.org/3/X9601E/x9601e04.htm.
3 Unethical practices in agriculture have been described by the Union of International Association in their Encyclopedia of World Problems and Human Potential: http://encyclopedia.uia.org/en.
4 In the United Kingdom, the Nuffield Council for Bioethics doing research on the ethical issues of developing technologies in agriculture. For example, Farming Tomorrow: The Genetic Technology (Precision Breeding) Act 2023, https://www.nuffieldbioethics.org/blog/the-future-of-farmed-food-in-england-what-has-the-genetic-technology-precision-breeding-act-2023-changed.
5 Genome editing and its use as a tool to develop new plant species present ethical and political challenges that have been little discussed until now. A recent report of the INRA-CIRAD-IFREMER Ethics Advisory Committee turns a lens on the intricacies of these questions, analyzing the value systems and symbolic representations that underlie the wide-ranging issues it identifies. Its discussions focus on the example of the

CRISPR-Cas9 genome-editing system. https://www.inrae.fr/en/news/ethics-and-politics-plant-genome-editing-joint-committee-report-and-position.

9 References

Arthur, W. B. (1989). Competing technologies, increasing returns, and lock-in by historical events, *The Economic Journal* 99(394), 116–113.

Barnes, B. and Dupre, J. (2008). *Genomes and What to Make of Them*, The University of Chicago Press, Chicago.

Blok, V. (2019). From participation to interruption: Toward an ethics of stakeholder engagement, participation and partnership in corporate social responsibility and responsible innovation. In: Von Schomberg, R. (Ed.). *International Handbook on Responsible Innovation. A Global Resource*, 243–257.

Blok, V. and Gremmen, B. (2016). Ecological innovation: Critical reflections on biomimicry as a new way of thinking and acting ecologically, *Journal of Agricultural and Environmental Ethics* 29(2), 203–217.

Blok, V. and Lemmens, P. (2015). Critical reflections on the concept of responsible innovation. In: Koops, B. J., van den Hooven, I. and Romijn, H. (Eds.). *Responsible Innovation: Issues in Conceptualization, Governance and Implementation*. Springer, Dordrecht, The Netherlands.

Boersma, R. and Gremmen, B. (2018). Genomics? That is probably GM! The impact a name can have on the interpretation of a technology, *Life Sciences, Society and Policy* 14(1), 8.

Boersma, R., Poortvliet, P. M. and Gremmen, B. (2019). The elephant in the room: How a technology's name affects its interpretation, *Public Understanding of Science* 28(2), 218–233.

Bos, A. P., Groot Koerkamp, P. W. G. G., Gosselink, J. M. J. and Bokma, S. (2009). Reflexive interactive design and its application in a project on sustainable dairy husbandry systems, *Outlook on Agriculture* 38(2), 137–145.

Brandt, K. and Barrangou, R. (2019). Applications of CRISPR technologies across the food supply chain, *Annual Review of Food Science and Technology* 10, 133–150.

Bruijnis, M. R. N., Blok, V., Stassen, E. N. and Gremmen, H. G. J. (2015). Moral 'lock-in' in responsible innovation: The ethical and social aspects of killing day-old chicks and its alternatives, *Journal of Agricultural and Environmental Ethics* 28(5), 939–960.

Carrillo-Hermosilla, J. (2013). Technological lock-in. Available at: http://www.eoearth.org/view/article/156453/. Accessed April 2022, The Encyclopedia of Earth.

Collingridge, D. (1980). *The Social Control of Technology*, St. Martin, New York.

Comstock, G. (2010). Ethics and genetically modified foods. In: Gottwald, F. T., Ingensiep, H. W. and Meinhardt, M. (Eds.). *Food Ethics*. New York, pp. 49–66.

Court of Justice of the European Union (CJEU) (2018). Judgment of the court from 25, *ECLI:EU:C:2018:583*. Available at: http://curia.europa.eu/juris/document/document.jsf?text=&docid=204387&pageIndex=0&doclang=EN&mode=req&dir=&occ=first&part=1&cid=44391. Accessed 8 January 2019, C-528/16.

David, P. A. (1985). Clio and the economics of QWERTY, *American Economic Review, Papers and Proceedings* 75, 332–337.

European Commission (2011). *Horizon 2020 – The Framework Programme for Research and Innovation*, Brussels.

European Parliament and Council of the European Union (2001). Directive 2001/18/EC of the European Parliament and of the council of 12. March 2001 on the deliberate release into the environment of genetically modified organisms and repealing Council Directive 90/220/EEC – commission Declaration. In: *Official Journal of European Communities*, L 106/44, 17.04.2001, 1–39.

FAO (2012). *The State of Food Insecurity in the World*, Food and Agriculture Organization, Rome, Italy.

Gaskell, G., Allansdottir, A., Allum, N., Castro, P., Esmer, Y., Fischler, C., Jackson, J., Kronberger, N., Hampel, J., Mejlgaard, N., Quintanilha, A., Rammer, A., Revuelta, G., Stares, S., Torgersen, H. and Wager, W. (2011). The 2010 Eurobarometer on the life sciences, *Nature Biotechnology* 29(2), 113–114.

Godin, B. (2015). *Innovation Contested: The Idea of Innovation Over the Centuries*, Routledge, New York.

Goubau, A. (2011). The AquAdvantage salmon controversy – A tale of aquaculture, genetically engineered fish and regulatory uncertainty. Available at: http://nrs.harvard.edu/urn_3HUL.Inst.Repos.8789564.

Gregorowius, D., Lindemann-Matthies, P. and Huppenbauer, M. (2012). Ethical discourse on the use of genetically modified crops: A review of academic publications in the fields of ecology and environmental ethics, *Journal of Agricultural and Environmental Ethics* 25(3), 265–293.

Gremmen, B. (2005). Genomics and the intrinsic value of plants, *Genomics, Society and Policy* 1(3), 1–7.

Gremmen, B. (2006). Plant genomics and the precautionary principle. In: Nap, J.-P., Atanassov, A. and Stiekema, W. (Eds.). *Genomics for Biosafety in Plant Biotechnology, NATO Science Series I: Life and Behavioural Sciences* (Vol. 359). IOS Press, pp. 223–233.

Gremmen, B. (2009). Biotechnology: Plants and animals. In: Olsen, J. K. B., Pedersen, S. A. and Hendricks, V. F. (Eds.). *A Companion to the Philosophy of Technology*, Wiley-Blackwell, 402–406.

Gremmen, B., Blok, V. and Bovenkerk, B. (2019). Responsible innovation for life: five challenges agriculture offers for responsible innovation in agriculture and food, and the necessity of an ethics of innovation, *Journal of Agricultural and Environmental Ethics* 32(5–6), 673–679.

Gremmen, B., Bruijnis, M. R. N., Blok, V. and Stassen, E. N. (2018). A public survey on handling male chicks in the Dutch egg sector, *Journal of Agricultural and Environmental Ethics* 31(1), 93–107.

Gremmen, H. G. J. (2007). De zwakste schakel: over maatschappelijk verantwoorde genomics, Inaugural Lecture, Wageningen University, Wageningen, The Netherlands.

Gremmen, H. G. J. (2017). Ethics of livestock farming? Who cares? Inaugural Lecture, Wageningen University, Wageningen, The Netherlands.

Grunwald, A. (2010). From speculative nanoethics to explorative philosophy of nanotechnology, *NanoEthics, Ethics for Technologies That Converge at the Nanoscale*, 2010.

Hanssen, L. and Gremmen, B. (2013). Influencing governance of a public-private partnership in plant genomics: The societal interface group as a new instrument for public involvement, *Public Understanding of Science* 22(6), 718–729.

Haperen, P. F. van, Gremmen, B. and Jacobs, J. G. M. (2012). Reconstruction of the ethical debate on naturalness in discussion about plant-biotechnology, *Journal of Agricultural and Environmental Ethics* 25(6), 797–812.

Holland, A. and Johnson, A. (1998). *Animal Biotechnology and Ethics*, Chapman and Hall, London, UK.

Jasanoff, S., Hurlbut, J. B. and Krishanu, S. (2015). CRISPR democracy: Gene editing and the -Need for inclusive deliberation, *Issues in Science and Technology* 32(1), 25–32.

Joint Research Centre (JRC) and Institute for Prospective Technological Studies (IPTS) (2011). *New Plant Breeding Techniques: State-of-the-Art and Prospects for Commercial Development*, Luxembourg.

Klerkx, L. and Rose, D. (2020). Dealing with the game-changing technologies of Agriculture 4.0: How do we manage diversity and responsibility in food system transition pathways?, *Global Food Security* 24, 1–7.

Leenstra, F., Munnichs, G., Beekman, V., Van den Heuvel-Vromans, E., Aramyan, L. and Woelders, H. (2010). Killing day-old chicks? Public opinion regarding potential alternatives, *Animal Welfare* 20, 37–45.

Levitt, M. (2003). Public consultation in bioethics: What's the point of asking the public when they have neither scientific nor ethical expertise?, *Health Care Analysis* 11(1), 15–25.

Locke, S. (2002). The public understanding of science – A rhetorical invention, *Science, Technology, and Human Values* 27(1), 87–111.

Meissner, D. and Carayannis, E. G. (2017). Value generation from industry-science linkages in light of targeted open innovation, *Journal of Knowledge Management* 21(2), 295–307.

Nekrasov, V., Wang, C., Win, J., Lanz, C., Weigel, D. and Kamoun, S. (2017). Rapid generation of a transgene-free powdery mildew resistant tomato by genome deletion, *Scientific Reports* 7(1), 482.

New Techniques Working Group (2008). Final report. Available at: http://www.seemneliit.ee/wp-content/uploads/2011/11/esa_12.0029.pdf. Accessed 19 June 2018.

Nuffield Council on Bioethics (2016). *Genome Editing: An Ethical Review*, London, UK.

Organisation for Economic Cooperation and Development (2009). *The Bioeconomy to 2030: Designing a Policy Agenda*, Main Findings, Paris, France.

Oudshoorn, N. E. J. and Pinch, T. J. (2003). How users and non-users matter. In: *How Users Matter: The Co-Construction of Technologies and Users*, MIT Press, Cambridge, MA, 4–22.

Perkins, R. (2003). Technological "lock-in", *Internet Encyclopaedia of Ecological Economics*, Accessed April 2015.

Poortvliet, P. M., Purnhagen, K. P., Boersma, R. and Gremmen, B. (2019). On the legal categorisation of new plant breeding technologies: Insights from communication science and ways forward, *European Journal of Risk Regulation* 10(1), 180–186.

Rip, A., Misa, T. J. and Schot, J. W. (Eds.). (1995). *Managing Technology in Society: The Approach of Constructive Technology Assessment*, Pinter Publishers, London, UK.

Rollin, B. E. (1995). *The Frankenstein Syndrome: Ethical and Social Issues in the Genetic Engineering of Animals*, Cambridge University Press, New York.

Rollin, B. E. (2006). *Science and Ethics*, Cambridge University Press, Cambridge, UK.

Sandoe, P. and Christiansen, S. B. (2008). *Ethics of Animal Use*, Wiley, New York.

Schomberg, V. and von, R. (2013). A vision of responsible research and innovation. In: Owen, R., Heintz, M. and Bessant, J. (Eds.). *Responsible Innovation*, John Wiley, London, UK.

Schomberg, V., von, R. and Hank, J. (2019). *International Handbook on Responsible Innovation: A Global Resource*, Edward Elgar Publishing, Cheltenham, UK.

Shu, Q. Y., Forster, B. P. and Nakagawa, H. (Eds.). (2011). *Plant Mutation Breeding and Biotechnology*, Food and Agriculture Organization, Rome, Italy.

Siipi, H. (2004). Naturalness in biological conservation, *Journal of Agricultural and Environmental Ethics* 17(6), 457–477.

Singer, P. (Ed.). (1986). *Applied Ethics*, Oxford University Press, Oxford, UK.

Sprink, T., Eriksson, D., Schiemann, J. and Hartung, F. (2016). Regulatory hurdles for genome editing: Process- vs. product-based approaches in different regulatory contexts, *Plant Cell Reports* 35(7), 1493–1506.

Stilgoe, J., Owen, R. and Macnaghten, P. (2013). Developing a framework for responsible innovation, *Research Policy* 42(9), 1568–1580.

Then, C. and Bauer-Panskus, A. (2017). *Playing Russian Roulette With Biodiversity*, Munich, Germany.

Thompson, P. B. (2007). *Food Biotechnology in Ethical Perspective*, Springer, Dordrecht, The Netherlands.

Van de Poel, I. and Robaey, Z. (2017). Safe-by-design: From safety to responsibility, *NanoEthics* 11(3), 297–306.

Veen, M., te Molder, H. F., Gremmen, B. and Van Woerkum, C. (2010). Quitting is not an option: An analysis of online diet talk between celiac disease patients, *Health* 14(1), 23–40.

Wathes, C. M., Kristensen, H. H., Aerts, J.-M. and Berckmans, D. (2008). Is precision livestock farming an engineer's daydream or nightmare, an animal's friend or foe, and a farmer's panacea or pitfall?, *Computers and Electronics in Agriculture* 64(1), 2–10.

Zambon, I., Cecchini, M., Egidi, G., Saporito, M. G. and Colantoni, A. (2019). Revolution 4.0: Industry vs. agriculture in a future development for SMEs, *Processes* 7(1), 36. https://doi.org/10.3390/pr7010036.

Zhang, H., Zhang, J., Wei, P., Zhang, B., Gou, F., Feng, Z., Mao, Y., Yang, L., Zhang, H., Xu, N. and Zhu, J. K. (2014). The CRISPR/Cas9 system produces specific and homozygous targeted gene editing in rice in one generation, *Plant Biotechnology Journal* 12(6), 797–807.

Chapter 12

Intellectual property rights in agriculture: ethical issues

Leland L. Glenna, The Pennsylvania State University, USA

1 Introduction

The expansion of intellectual property (IP) rights, particularly utility patents, in United States (US) agriculture over the past 50 years is grounded in two social phenomena. The first is the development of agricultural biotechnologies. The second is the development of the knowledge economy. Agricultural biotechnology refers to the use of molecular biological techniques ostensibly to solve agricultural problems, such as the use of genetic engineering to produce novel crop traits (Fernandez-Cornejo, 2004). The knowledge economy refers to a shift in society toward a greater reliance on knowledge, information, and more sophisticated skills and technologies to produce goods and services. The justification for its development is to direct more economic rewards to people who invest in and produce the knowledge and information that become additional raw materials in productive activity (OECD, 1996). After utility patents became prominent for US agricultural crops, they spread to other parts of the world and became a key dimension of global trade policy.

http://dx.doi.org/10.19103/AS.2023.0125.13

Knowledge production was not always seen as something that could or should be rewarded through privatization. In the 1950s, prominent economists like Samuelson (1954, 1955) argued that government expenditures were necessary to promote knowledge production because knowledge had qualities that made it difficult to privatize. For example, unlike a piece of equipment, knowledge can be shared freely and it is less than easily excludable. After all, people can share knowledge through conversations. Knowledge also tends to be non-rivalrous, which means that applying knowledge does not prevent others from using it. Because of these non-excludable and non-rivalrous qualities, private firms and individual investors cannot profit from the creation of knowledge without intellectual property and knowledge-transfer policies that facilitate returns on investments for knowledge production.

One important example of such a policy change is the 1980 Bayh-Dole Act, which enabled the patenting of publicly funded university research so that it could then be licensed to the private sector (Eisenberg, 1996). After the US adopted the Bayh-Dole Act, many other industrialized nations followed with similar legislation (Gering and Schmoch, 2003). That policy reasoning became institutionalized globally through the Global Agreement on Tariffs and Trade and the World Trade Organization (Mirowski, 2011). Agricultural biotechnology became an important mechanism through which agricultural knowledge, especially university-based agricultural research, could become privatized, making it attractive to private research and development investors (Brandl and Glenna, 2017).

Because the convergence of biotechnology and the knowledge economy are especially evident in the agricultural crop seed sector, I will focus on crop seeds in this chapter. Before doing that, however, I will briefly explain the reasoning behind the promotion of IP protection. I will then describe the effects of these changes on the crop seed sector and highlight several ethical issues that have emerged because of these changes.

2 Justification for intellectual property protection

IP includes patents, trademarks, copyrights, and trade secrets. Each type of IP is used in agriculture and food research and development, but the extension of patents to crop seeds is the most obvious example of how IP applications have expanded and become stricter.

Patents have been around for centuries. There are several examples of legal protections for inventions and innovations in Ancient Greece and Rome. However, according to historians, 'the first real patents' can be traced to Northern Italy in the fifteenth century, and many European countries subsequently developed some form of protection for inventions (Frumkin, 1945: 145). Despite these early examples, the contemporary legal framework for patents emerged in the US and France in the late eighteenth century. According to Lessig (2004: 4), the original intent behind US patent policies, established by Congress in 1790, was to 'promote the progress of science . . . by securing for limited times exclusive rights' (Lessig, 2004: 4). The US tradition of granting patents to inventors is to provide monopoly protection for a specified period as an incentive to develop the patent and to promote its development and application. Without that patent protection, inventors and investors might not contribute time and resources to develop the invention. Furthermore, without a patent, inventors might strive to keep the knowledge behind their inventions secret. With the patent the inventor is given the incentive to invest in innovative research because they are granted exclusive rights to profit from the invention even after making the information public since that public knowledge might stimulate more innovation in the future (Boldrin and Levine, 2013: 2; Lei et al., 2009).

Although patenting has existed since the beginning of US history, some have begun to question whether the recent expansion in scope for patent protection, as well as the court decisions that have enhanced the power of the patent holders, has disrupted the balance between the goal of providing incentives for individuals and the goal of extending and stimulating future innovations. Specifically, scholars question whether patent rights have become so restrictive that they, ironically, are now limiting innovations and social progress (Baker et al., 2017; Glenna et al., 2015; Boldrin and Levine, 2013; Lessig, 2004; Jaffe and Lerner, 2004). If patents are not stimulating innovation, the justification for the expansion and intensification of patents collapses.

3 Changes in the crop seed sector

Heald and Chapman (2012) examine the changes in the availability of plant varieties during the twentieth century to determine if the introduction of IP reduced or expanded the number of varieties. As they point out, if economic theory is correct, the introduction of the 1930 Plant Patent Act, the passage of the 1970 Plant Variety Protection Act, and the acceptance of utility patents on crops by the US Patent Office in 1985 should have stimulated the production of new plant varieties. By contrast, some social scientists and ethnobotanists argue that IP protection promotes monocultures and reduces genetic and varietal diversity. After reviewing the data, Heald and Chapman (2012) discover that neither argument is quite right. First, they found that varietal diversity increased. Second, they found that IP policies 'played little or no role in that innovation' (Heald and Chapman, 2012: 1056). Furthermore, they contend that increased IP protection stimulates rent-seeking more than it stimulates innovation. Looking specifically at growth in tomato varieties, which have few patents, but many varieties, they conclude, 'Clearly, innovation does not require a patent system for tomatoes' (Heald and Chapman, 2012: 1097).

Although Heald and Chapman's (2012) paper is an important study because it undermines the justification for applying patents to agricultural crops, it tends to treat the 1930 Plant Patent Act, the 1970 Plant Variety Protection Act, and the 1985 application of utility patents as similar types of IP. This is a problem because many scholars highlight important distinctions among them, especially between patents and the earlier types of IP.

Crop breeders and seed companies were not allowed to secure utility patents on living organisms prior to 1980, but policymakers created mechanisms that enabled them to secure some return on research investments. In 1961, an intergovernmental organization headquartered in Geneva, Switzerland, established the International Union for the Protection of New Varieties of Plants. It became widely known by its acronym UPOV, based on the French expression of the International Union for the Protection of New Varieties of Plants. The US passed the Plant Variety Protection Act (PVPA) in 1970, which was similar to UPOV. The US then joined the UPOV in 1980. UPOV and PVPA allow farmers to save seed, as long as farmers are saving seed for replanting, and not for sale. And university scientists face no research restrictions under UPOV or PVPA. Therefore, this legal framework is less restrictive for farmers and researchers than utility patents (Huffman and Evenson, 2006; Buttel and Belsky, 1987).

Crop breeding companies also found other ways to secure returns on their investments. Although they were achieving improved crop yields in maize by developing both open-pollinated and hybrid varieties, the United States Department of Agriculture (USDA) began promoting hybridization because it would create incentives for the coveted private-sector investments. Because yields are not passed to the second generation in hybrid maize, it is not feasible for farmers to save hybrid maize seeds for planting the next season. That means that farmers must return to crop breeding companies to get the improved hybrid seeds each year. Kloppenburg (2004: 319), therefore, refers to hybridization as a 'biological patent', because the private sector recognized it as a mechanism for gaining a return on investment.

A watershed moment occurred in 1980 when the US Supreme Court decision, Diamond v. Chakrabarty, allowed inventors to secure patents on living organisms, and the US Patent Office extended this to plants in 1985. Prior to this decision, utility patents were not applicable to living organisms like plants. Then, as I pointed out earlier, this decision eventually became global policy through the Trade-related Aspects of Intellectual Property Rights, which standardized IP rights through the World Trade Organization (Strauss, 2009). Private firms exercised this new patenting ability through biotechnological tools used to develop transgenic plants.

In addition to expanding the scope of patents to include living organisms, courts facilitated the expansion of those rights, even when elements of the legal, legislative, and regulatory decisions and policies seemed contradictory. For example, Pechlaner (2012) describes how firms like Monsanto managed to use their vast legal resources to privatize the benefits of transgenic seeds, while socializing the costs, in the US and Canada, where around half of the world's transgenic crops are grown. For example, Monsanto was given the ability to force farmers to pay compensation if those farmers replanted any of the seeds from the previous year's harvest. However, if any of the genetic material in the seeds were to spread from farmers' fields to their neighbors' fields, the farmers would be held responsible, not the firm. Therefore, Monsanto was granted property rights to the genetic material when Monsanto could benefit economically, but Monsanto's property rights did not hold it accountable for the spread of that genetic material when that spread caused harm.

In 2001, the US Supreme Court again made a monumental decision in Pioneer Hi-Bred v. J.E.M Ag Supply, which extended utility patents to conventionally bred crops for the first time (Janis and Kesan, 2002; Sease, 2007). Because utility patents are more restrictive than PVPA and UPOV, 'it is not surprising that in the USA breeders are massively applying for patent protection for their varieties' (Le Buanec and Ricroch, 2014: 69).

Although these policies often became globalized, there was still geographical variation in their application. For example, the European Union (EU) made it possible for farmers to save seed under some circumstances but also made it possible for crop breeders to charge a fee to farmers who saved seed. This may seem restrictive, but it is important to keep in mind that that fee is lower than the cost of purchasing the seeds from breeding companies (Galushko and Gray, 2014). Therefore, EU policies did not provide the same incentives to rent-seeking and market consolidation that US policies did.

Furthermore, the policy changes altered the relationships between universities and seed companies and farmers. Because a substantial amount of the research behind agriculture and food innovations in the US occurs in universities, the authors of these policy changes sought to include universities in the knowledge economy. Combined with the 1980 Bayh-Dole Act, which enabled universities to patent those inventions and to license them to the private sector, this constellation of policies created the institutional framework for privatizing previously public university agricultural research. Therefore, although IP for agricultural crops was not new, the use of utility patents was new, and that marked an important transition. Until the 1980s, much of the agricultural seed sector could be characterized as being comprised of numerous small seed companies that worked closely with land-grant universities and land-grant experiment stations to facilitate the transfer of improved seed varieties from agricultural scientists to farmers (Fernandez-Cornejo, 2004). The market became much more concentrated after patents were introduced (Fuglie et al., 2012).

4 Ethical considerations

There are numerous ethical considerations related to the application of patents to agricultural crops. One of the most basic questions is an instrumental one. Proponents of science commercialization argue that social welfare is enhanced through the efficient transfer of research from the public to the private sector (Etzkowitz, 2001; Link and Siegel, 2007). Some counter that societies need a combination of private and public goods and that the focus on commercialization is undermining the broader benefits that come from public science (Dasgupta and David, 1994). Despite the tendencies toward market concentration, the loss of focus on public goods, and the loss of interest in the broad benefits of public research, a Utilitarian might still argue that these downsides are acceptable because the policy changes maximize utility and enhance social welfare.

4.1 Social welfare

Measuring social welfare contributions is complex, in part because not everyone agrees on how to define or measure contributions to social welfare. Economists often evaluate the contributions of people, capital and other resources through production-function models. However, economists using production-function models have failed to demonstrate that scientific research contributes to economic growth (Mirowski, 2011). This has led some to argue that science should not be thought of as a production input, but rather as part of the broader infrastructure needed to enable innovations to happen (Mazzucato, 2013; Mirowski, 2011). My point is not to resolve the debate on how to conceptualize the contributions of science to social well-being. My point is to clarify that trying to evaluate the Utilitarian claims about the contributions of science to society is complex. Nevertheless, there are ways to evaluate whether more patenting in the agricultural crop sector has been beneficial.

One of the main justifications for the development of the knowledge economy is that patenting knowledge will attract more private investment in scientific research and development and yield more productivity. In general, however, evidence generally does not support these claims (Mirowski, 2011). In the agriculture and food sector, private firms have substantially increased investments in research and development, but most of that research has focused on research that enhances company profits, while investments in public research have declined (Glenna, 2017). Moreover, because the agriculture and food system has long relied on public-goods research conducted in public research institutions to drive innovation, many have raised concerns about the capacity to sustain socially beneficial research and development (Galushko and Gray, 2014; Fuglie et al., 2012; Piesse and Thirtle, 2010; Huffman and Evenson, 2006). One of the most notable examples of the decline in social welfare can be captured in the slowing of and even decline in crop yield increases as privatization has increased.

I do not want to treat crop yields as a perfect measure of social welfare. Yield increases have been linked to declines in farm numbers, a phenomenon referred to as the 'technology treadmill', whereby early adopters gain the economic benefits while later adopters, who are unable to keep up, fall off the treadmill (Levins and Cochrane, 1996). Therefore, not all segments of society benefit from yield increases. However, economists, agricultural scientists, and policymakers have tended to think of increasing crop yields as a measure of social welfare because there is an assumption that consumers will be able to spend less to satisfy their food preferences (Fuglie et al., 2012).

I earlier noted that Heald and Chapman (2012) treated all plant IP the same, which is problematic. Nevertheless, their analysis is useful because they compare field trial information on patented corn varieties from 1986 to 2005, which would include the era in which utility patents were applied to corn crops. They find that patented hybrid varieties tended to produce lower yields than non-patented hybrids. Since yield increases are often treated as a public good, the findings suggest that patenting crops do not generate social welfare.

Other research indicates that crop yield increases have also slowed in the US since 1980 when the Bayh-Dole Act was passed and when the US Supreme Court decided that patents could be applied to living organisms. For example, crop yield increases vary by crop in the US, as well as by nation. In the US, patents were common in genetically engineered crops, like maize and soybeans, but have not yet been widely applied in crops like wheat, oats and barley since they were not genetically engineered. Since patented crops attract more private investments, it is not surprising that those yields for patented crops increased at a faster rate than non-patented crops (Brandl and Glenna, 2017). One might conclude from this finding that patent protections should be enhanced and expanded to more crops so that more private investments will follow. However, such a conclusion would be off the mark because wheat, oats, and barley have shown a steady yield increase in countries that do not have transgenic crops (e.g. Germany) (Brandl and Glenna, 2017). Furthermore, research indicates that privatizing crop research has not had desired outcomes. The United Kingdom (UK) is an interesting case study because the UK sought to privatize crop research so that the UK could then reduce public expenditures on crop research. After the privatization of wheat research, wheat yield growth slowed substantially. Having studied the slowdown, Galushko and Gray (2014) conclude that one contributing factor is that privatization reduced investments in the basic research necessary to improve seed quality. To put it bluntly, crop yields have improved over the last century because governments have invested in scientists and public research facilities to conduct basic research on genetics, plant physiology, plant pathology and other research that provides the foundation for crop improvements.

Perhaps even more challenging to the notion that patents enhance innovation, research indicates that patents may be hindering public research. Analyzing a survey of entomologists, Glenna et al. (2015) explain that patents are used by private firms to limit the research that many university scientists think is necessary to adequately evaluate efficacy and environmental risks. Those patent restrictions also limit scientific research that is necessary for future innovations. Glenna et al. (2015) conclude that the evidence suggests that risk assessments and innovations are being compromised to protect short-term IP interests of private firms.

Another negative outcome of the rise in patenting in crops is that it has facilitated market concentration in the seed sector. Soon after utility patents for crops were introduced, a few private firms came to dominate the seed market. In the US, by the early 2000s, the three largest seed firms held 85% of genetically engineered corn patents and 69.6% of non-corn, genetically engineered plant patents (Glenna and Cahoy 2009). Globally, in 1985, the market share of the nine largest firms was 12.7% of the seed market. By 1996, this share increased to 16.7%. In 2009, just three of the largest companies (Monsanto, DuPont, and Syngenta) had a market share of 34% (Schenkelaars et al., 2011). When buyers or sellers become concentrated in a market, they tend to gain power to set prices. In the case of seeds, when farmers have only a few companies from whom to buy seeds, those companies no longer have to compete to improve products or lower prices.

It is important to recognize that the level of concentration varies by nation. In Germany, which does not have genetically engineered crops and where IP policies do not allow for patenting and exclusive licenses to the same extent they do in the US, there is less market concentration (Brandl and Glenna, 2017). Moreover, Germany's yield increases for crops not genetically engineered or patented, including wheat and barley, outperformed those same crops in the US. Other research offers similar conclusions. Research on wheat indicates that IP protections have no discernable effect on promoting higher-yielding varieties. Instead, they tend to promote rent-seeking behaviors, such as the pursuit of patents to gain rights of ownership, but little investment in research and development of those crops (Heald and Chapman, 2012; Alston and Venner, 2002). This is consistent with findings in other crops, including tobacco, soybeans, and canola (Heald and Chapman, 2012).

Based on the comparison of yields and market concentration in the US and Germany, Brandl and Glenna (2017) conclude that the commercialization approach, which includes more expansive and stricter patenting policies, generates less social welfare than an approach that focuses on public investments to promote the generation of public goods. They contend that IP policies should be weakened to promote more innovation and problem-solving. Patented crops have certainly introduced some social benefits. For example, private firms might not have invested in the research and development necessary to generate crops that produce their own pesticides, which have led to reductions in some types of environmentally harmful agricultural chemicals (NAS, 2016). However, a counter example is the virus-resistant papaya developed by the public sector (NAS, 2016). Furthermore, research on recent efforts to harness private investments for high-technology solutions to agricultural problems may simply crowd out more accessible and more workable solutions (Fairbairn et al., 2022). In sum, evidence does not support claims that patenting is necessary to enhance social welfare. In fact, some evidence suggests that it impedes social welfare.

5 Other ethical concerns

Utilitarianism is not the only way to evaluate whether stricter IP policies in agriculture are a good thing. Even if there were clear evidence that stricter IP policies were necessary to promote innovation and if the innovation yielded more goods for more people, there might still be ethical reasons not to do it. One of the most basic questions is whether any living organism or part of that organism should be patented. I noted earlier that the 1980 US Supreme Court decision *Diamond v. Chakrabarty* laid the groundwork for patenting living organisms. Chakrabarty, an engineer at General Electric, manipulated a bacterium to get it to incorporate some additional plasmids. However, the US Patent Office ruled that '"products of nature" that are "living organisms" cannot be patented' (Stix, 2006: 78). When the appeal of this ruling reached the US Supreme Court, the biotechnology industry was just emerging. Economic interests combined with a greater scientific and engineering expertise to support an argument for patenting. For example, Cohen and Boyers DNA modification, which laid the foundation for biotechnology, had been patented just a few years before the Chakrabarty decision (Stix, 2006).

Although patents on living organisms or elements of those living organisms have become common, ethical concerns remain. Some question the morality of patenting living organisms. Some nations, including Japan and the EU, prohibit patents on 'inventions liable to contravene public order, morality or public health' (Chambers, 2002: 226). By contrast, Chambers (2002) observes that the US seldom considers the morality of a patent application and has generally been more favorable toward patenting living organisms, as long as they are considered inventions. However, even the US Patent and

Trademark Office raised morality concerns when an applicant sought a patent for a human/nonhuman chimera (Chambers, 2002). Moreover, in 2013, the US Supreme Court ruled that one cannot gain patents on human genes, even if the entity seeking the patent successfully isolated the gene (Liptak, 2013). Therefore, even the US establishes moral boundaries occasionally.

Other ethical concerns arise less over the underlying morality and more over the political/economic implication of patenting, including biopiracy. In 2001, many people in the nation of India were outraged when a US firm was granted patents to three strains of basmati rice. Just because a firm had created new varieties by crossing basmati rice seed with American long-grain rice seed, some argued, did not justify a patent when Indian farmers had been growing basmati rice 'for centuries in the foothills of the Himalayas' (Harding, 2001). In another case, a US firm was granted a patent on a Mexican yellow bean variety. The patent gave the firm monopoly control, which enabled it to exclude farmers from Central America and Mexico from selling their yellow beans in the US. However, scientists evaluating the genome of the patented yellow bean concluded that it had much in common with an existing Mexican cultivar. They therefore questioned why it had received a utility patent or a plant variety protection certificate (Pallottini et al., 2004).

6 Positive versus native rights

Any justification for utility patents is premised on the distribution of property rights within a market framework. Therefore, it is important to explore how the distribution of property rights is presumed to work. Many controversies related to agriculture and food biotechnologies can be understood as conflicts over the application of negative rights (i.e. IP rights that protect a person from interference by others) as opposed to positive rights, which establishes a duty by others to provide a good or service (Hirschl, 2000).

As I have already explained, greater sophistication in biotechnologies, from transgenic engineering to gene editing, has been accompanied by more restrictive IP protections, which prohibit others from making, using or selling the invention without authorization. This means that negative rights are becoming stricter. In contrast, older forms of plant modification, such as cross-breeding and radiation or chemical mutagenesis, were eligible for less restrictive protections, such as through the USA's Plant Variety Protection Act or the international UPOV, which allow farmers to save part of the crop for replanting the next season and allow breeders to build innovations on existing crops. Some have focused attention on how the food system could be built on a positive-rights approach (Anderson, 2008).

Developing agricultural biotechnology in the context of positive rights could alter the nature of biotechnology controversies. For example, an approach based on peoples' right to food could offer the foundation for reforming IP laws surrounding agricultural biotechnology applications. Under a positive-rights regime, the research and development process of agricultural innovations would be required to direct innovations toward enhancing the food supply, rather than using patents to prevent farmers from saving and replanting seeds or preventing crop breeders from developing and distributing new crops from an existing variety. Since emerging agricultural biotechnologies would then be applied to solving environmental and agronomic problems with less concern for exclusions, some of the resistance to these technologies might be reduced.

7 The ethical limits of markets

Another ethical concern over the use of utility patents in agriculture is the implications and consequences of how combining agricultural biotechnology and the knowledge economy fostered the commercialization of science. In the early 2000s, a university administrator explained, 'By definition, the best commercial interest means that the public is buying it, which means that the public is benefiting. A measure of public good is the moving of a commercial product' (Glenna et al., 2007: 155). Another university administrator said that universities contribute to the public good when they 'create a capacity within the state for economic growth', and one explained, 'Let me put it this way, if we didn't do it [generate IP], we would receive less funding. It could have a negative effect [on our funding] if we weren't transferring our research over to private enterprise' (Glenna et al., 2007: 156). The comments elevate concerns raised by Sandel (2012) about the tendency over the past few decades to rely increasingly on markets to determine the value of ever more spheres of our lives. As a thought experiment, he asks if anything is lost when children are paid to read books, when people are paid for their organs, or when an auction is used to determine who will win an award, win a national athletic event, be admitted to elite universities, or be granted US citizenship. The point of his ethical reasoning is to question if there are parts of our lives that should not be evaluated by the price they can get in a market. Most reasonable people are likely to agree that Sandel (2012) succeeds in demonstrating that society is worse off when markets replace many nonmarket norms. Children paid to read books may never learn to appreciate the intrinsic value of reading. Auctioning off an award corrupts the purpose of a practice or event. These insights can be applied to the expansion and intensification of patents in agriculture.

8 Undermining institutional integrity: the university

Somers and Block (2005) refers to the 1980 Bayh-Dole Act, the Chakrabarty decision, and subsequent policies as a 'market-fundamentalist model of governance', which is another way of describing what Sandel (2012) referred to as replacing nonmarket norms with market values. When these changes were still in their infancy, Walzer (1983) raised concerns about how nonmarket institutions would be corrupted if their organizing principles were replaced with market principles. He explained that societies need institutions that serve purposes beyond merely enhancing market exchanges and economic growth. Applying Walzer's (1983) insights to the realm of national research agendas, Block (2011) explains that, after the Second World War, the US and many other industrializing nations developed a welfare-state model of innovation when their governments invested public money in research with the goal of enhancing innovation. As I have already explained, applying utility patents and other types of IP to publicly funded research became a central feature of the knowledge economy. Block (2011) acknowledges that, although the government, the market economy, science and the university are all different spheres of society, they are also interconnected institutions in modern, industrial society. And he concedes that nations may indeed seek to enhance interactions between these institutions to promote innovations. However, he warns those efforts, if not done carefully, could undermine the public-interest principles of social institutions, including public research universities.

9 Undermining institutional integrity: science

Merton (1973) described science as an institution that operates according to nonmarket organizing principles. Most notably, science follows the principles of communalism, universalism, disinterestedness and organized skepticism. Although some have critiqued Merton's structural-functionalist framing of science, scholars have consistently found that public science has different values and goals than private science. Lacy (2001) summarized that university science tends to have a long-term perspective directed at advancing basic knowledge creation, societal responsibilities, problem-solving and open communication. By contrast, scientists working for private firms are expected to be proprietary, focus on producing marketable products that will generate a profit for their firm quickly, and be secretive.

As utility patents became more common in universities, many observers grew concerned that university science's values and goals would become more aligned with industry values and goals (Dasgupta and David, 1994; Powell and Owen-Smith, 1998; Slaughter and Rhoades, 2004). Analyzing a survey of scientists at private, public non-land-grant and land-grant universities, Glenna et al. (2011) found that some university scientists have a market orientation that is more aligned with industry scientists and that, when combined with industry funding, they are more likely to pursue IP protections for their research outputs and to conduct more applied research. Glenna et al. (2011) conclude that public research funds will need to be increased to maintain the university's commitment to basic knowledge and public goods.

Unfortunately, the trend in funding research is going in the opposite direction. The percentage of university research funding from the federal government has declined steadily over the past five decades. Meanwhile, university funding for research that comes from their own investments, including patenting research outputs, has steadily increased. This suggests that universities are seeking to enhance returns on research investments so that they can continue to support research at their universities (Glenna, 2017).

In public institutions, this corruption of purpose is observable in tangible ways. For example, Dickenson (2013) explains how the emergence of biotechnologies and the knowledge economy have affected public health. In the book *Me Medicine vs. We Medicine*, she explains that We Medicine refers to public-health measures, such as vaccinations, sanitation infrastructure, and food-safety regulations, that are meant to promote the health of all members of society. By contrast, Me Medicine focuses on research directed at developing products that individuals can purchase to address personal health issues. Marks (2014) applies a similar argument to the agriculture and food sector when he explains that, when research is directed at a return on investment, researchers will focus on generating a product for sale, rather than altering the system to promote more healthful foods. He offers the example of weight loss and explains that a market-based solution would focus on producing a weight-loss drug, rather than research on how diet and exercise can improve health or how the corporate-driven, high-calorie agriculture and food system might be the root cause of the problem.

10 Contributive justice and problem-solving

Distributive justice refers to the idea that the risks and benefits of a productive economy and access to social goods should be distributed fairly. Sandel (2021) acknowledges the importance of distributive justice debates, but he also argues that these debates tend to overemphasize people's consumption. He contends that the tradition of contributive justice has been largely absent from US political discourse since the 1960s, when Robert Kennedy emphasized how making a contribution can help people feel a sense of commitment to their communities, families and nation. Kennedy once stated that it is important for people to be able to say 'I helped build this country. I am a participant in its great public ventures' (quoted in Sandel, 2021: 212).

Without referring to the concept of contributive justice, Wright (2012: 1716) nevertheless addresses the concept when he discusses 'three related examples of mission-oriented agricultural innovations' during the twentieth century. The three examples are (1) the establishment of the land-grant-university system (which began in the latter part of the nineteenth century), (2) the intensive focus on enhancing agricultural crop yields in US agriculture, and (3) the intensive focus on extending those crop-yield improvements to other parts of the world through the Green Revolution in the 1960s and early 1970s. The land-grant system and the Green Revolution institutions still exist, but Wright (2012) bemoans that there have been no comparable grand missions in the past several decades and he worries that, because of the changes to agricultural and food research institutions in the US and the International Research Centers created by the Green Revolution effort, the US and the world will lack the institutional capacity to address future agriculture and food crises. Critics might take issue with Wright's (2012) overly positive portrayal of yield increases and the benefits of the Green Revolution. For example, some have pointed out that these yield increases often come at the expense of environmental challenges and declining farm numbers (Evenson and Gollin, 2003). However, that should not distract us from his fundamental concerns about how the lack of common purposes embodied in grand missions may be limiting our social cohesion and the ability of universities and governments to use science and technology to enhance the common good.

Great missions worked in the past because many university scientists and engineers enter their fields of study for reasons beyond financial returns, and society and universities once harnessed those extra-market motivations. Promoting research based on contributive justice could once again give scientists a sense of purpose that would motivate them to strive to solve social problems beyond the promise of financial returns for themselves or their universities. Of course, scientists want to earn a good living to support themselves and their families. But their primary motivation is often not to maximize economic returns from their labor. Rather, they are motivated to make a significant breakthrough, to solve a major problem, or even just to solve a puzzle. A research system based on contributive justice would present a direct challenge to the justification behind utility patents, as well as to the broader knowledge economy that is integral to the justification for utility patents.

11 Conclusion

Although utility patents have been awarded to inventors since the very beginning of the US as a nation, they were not applied to agricultural crops until the middle of the 1980s. The US and some other industrialized nations began applying other forms of IP protections to agricultural crops as early as the first half of the twentieth century, but these protections were not as strict as utility patents. They allowed farmers to save seeds and breeders to innovate on existing crops. That changed with the emergence of the knowledge economy and agricultural biotechnologies, two phenomena that were mutually reinforcing.

The knowledge economy reflected a shift in social governance that gave priority to market mechanisms, and agricultural biotechnologies fit within that knowledge economy as research-intensive products that could be patented more readily than agroecological approaches to farming. The most prominent justification for the application of patents to agricultural crops tends to rest on a combination of Utilitarian and negative-rights arguments. That justification is that more people will benefit from more inventions if the inventors are granted monopoly rights, which will inspire them to invest more time and money with the promise of a big payoff.

Empirical studies challenge the notion that social welfare is enhanced through the application of utility patents to agricultural crops. For example, crop yields in many crops were increasing at a greater rate when crops were treated as public goods than they have since they became patented. In fact, taking a positive-rights approach, by declaring that people have a right to food, for example, may lead to more equitable distributions of food than assigning patents to agricultural crops.

Other ethical concerns are also worth considering. The further privatization of agricultural knowledge can corrupt the public research institutions, such as agricultural universities, as well as the people who work in them. Patents privilege commodified solutions to problems, rather than solutions that involve collective action. The idea that patents are needed to attract people's commitment to solving social problems also tends to diminish how people interact with their societies. Rather than being motivated to contribute to a collective effort to solve problems, people are reduced to being motivated primarily to seek a profit for their efforts.

The use of patents in agriculture has not generated the social welfare benefits promised by proponents. There is even evidence that patents have undermined social welfare goals. In addition, there is evidence that social institutions have become corrupted and that the ethical dimensions of the global agriculture and food system have been diminished. Given the failure of patents to meet social welfare expectations, along with the other ethical concerns, policymakers should reconsider the use of utility patents for agricultural crops. Other approaches to structuring the institutions that govern agricultural research and development may generate more social welfare and may even enhance the humanity of the scientists and others engaged in the research and development.

12 References

Alston, J. M. and Venner, R. J. 2002. The effects of the US Plant Variety Protection Act on wheat genetic improvement. *Research Policy* 31(4): 527–542.

Anderson, M. D. 2008. Rights-based food systems and the goals of food systems reform. *Agriculture and Human Values* 25(4): 593–608.

Baker, D., Jayadev, A. and Stiglitz, J. 2017. Innovation, intellectual property, and development: A better set of approaches for the 21st century. Paper published by accessibsa and the Shuttleworth Foundation.

Block, F. 2011. Innovation and the invisible hand of government. In: Block, F. and Keller, M. R. (Eds), *State of Innovation: The U.S. Government's Role in Technology Development*, pp. 1–26. Boulder, CO: Paradigm Publishers.

Boldrin, M. and Levine, D. K. 2013. The case against patents. Working Paper No. 2012-035A. St. Louis, MO: Federal Reserve Bank of St. Louis.

Brandl, B. and Glenna, L. L. 2017. Intellectual property and agricultural science and innovation in Germany and the United States. *Science, Technology, and Human Values* 42(4): 622–656.

Buttel, F. H. and Belsky, J. 1987. Biotechnology, plant breeding, and intellectual property: Social and ethical dimensions. *Science, Technology, and Human Values* 12(1): 31–49.

Chambers, J. 2002. Patent eligibility of biotechnical inventions in the United States, Europe, and Japan: How much patent policy is public policy? *George Washington International Law Review* 34(1): 223–246.

Dasgupta, P. and David, P. A. 1994. Towards a new economics of science. *Policy Research* 23(5): 487–521.

Dickenson, D. 2013. *Medicine vs. We Medicine: Reclaiming Biotechnology for the Common Good*. New York, NY: Columbia University Press.

Eisenberg, R. S. 1996. Public Research and private development: Patents and technology transfer in government-sponsored research. *Virginia Law Review* 82(8): 1663–1725.

Etzkowitz, H. 2001. Beyond the endless frontier: From the land grant to the entrepreneurial university. In: Wolf, S. and Zilberman, D. (Eds), *Knowledge Generation and Technical Change: Institutional Innovation in Agriculture*, pp. 3–26. Boston, MA: Kluwer Academic Publishers.

Evenson, R. E. and Gollin, D. 2003. Assessing the impact of the green revolution, 1960 to 2000. *Science* 300(5620): 758–762.

Fairbairn, M., Kish, Z. and Guthman, J. 2022. Pitching agri-food tech: Performativity and non-disruptive disruption in Silicon Valley. *Journal of Cultural Economy* 15(5): 652-670 doi: 10.1080/17530350.2022.2085142.

Fernandez-Cornejo, J. 2004. The seed industry in U.S. agriculture: An exploration of data and information on crop seed markets, regulation, industry structure, and research and development. Agriculture Information Bulletin No. (AIB786). Washington, DC: United States Department of Agriculture, Economic Research Service.

Frumkin, M. 1945. The origin of patents. *Journal of the Patent Office Society* 27(3): 143–149.

Fuglie, K., Heisey, P., King, J., Pray, C. E. and Schimmelpfennig, D. 2012. Research funding. The contribution of private industry to agricultural innovation. *Science* 338(6110): 1031–1032.

Galushko, V. and Gray, R. 2014. Twenty five years of private wheat breeding in the UK: Lessons for other countries. *Science and Public Policy* 41(6): 765–779.

Gering, T. and Schmoch, U. 2003. Management of intellectual assets by German Public research organizations. In: *Turning Science into Business: Patenting and Licensing at Public Research Organisations*, pp. 169–188. Paris, France: OECD Publishing.

Glenna, L. L. 2017. AFHVS 2017 presidential address: The purpose-driven university: The role of university Research in the era of science commercialization. *Agriculture and Human Values* 34(4): 1021–1031.

Glenna, L. L. and Cahoy, D. R. 2009. Agribusiness concentration, intellectual property, and the prospects for rural economic benefits from the emerging biofuel economy. *Southern Rural Sociology* 24(2): 111–129.

Glenna, L. L., Lacy, W. B., Welsh, R. and Biscotti, D. 2007. University administrators, agricultural biotechnology, and academic capitalism: Defining the public Good to promote university-industry relationships. *The Sociological Quarterly* 48(1): 141–163.

Glenna, L. L., Tooker, J., Welsh, J. R. and Ervin, D. 2015. Intellectual property, scientific independence, and the efficacy and environmental impacts of genetically engineered crops. *Rural Sociology* 80(2): 147–172.

Glenna, L. L., Welsh, R., Ervin, D., Lacy, W. B. and Biscotti, D. 2011. Commercial science, scientists' values, and university biotechnology research agendas. *Research Policy* 40(7): 957–968.

Harding, L. 2001. India outraged as US company wins patents on rice. *The Guardian*. Available at: https://www.theguardian.com/world/2001/aug/23/highereducation.biologicalscience (Accessed 13 August 2022).

Heald, P. J. and Chapman, S. 2012. Veggie tales: Pernicious myths about patents, innovation, and crop diversity in the twentieth Century. *SSRN Electronic Journal* 2012(4): 1051–1102.

Hirschl, R. 2000. Negative. *Rights VS*. "Positive" entitlements: Comparative study of judicial interpretations of rights in an emerging neo-liberal economic order. *Human Rights Quarterly* 22(4): 1060–1098.

Huffman, W. E. and Evenson, R. E. 2006. *Science for Agriculture: A Long-Term Perspective*. Ames, IA: Blackwell Publishing.

Jaffe, A. B. and Lerner, J. 2004. *Innovation and Its Discontents: How Our Broken Patent System Is Endangering Innovation and Progress, and What to Do about It.* Princeton, NJ: Princeton University Press.

Janis, M. D. and Kesan, J. P. 2002. Intellectual property protection for plant innovation: unresolved issues after J.E.M. v. pioneer. *Nature Biotechnology* 20(11): 1161–1164.

Kloppenburg, J. R., Jr. 2004. First the seed: The political economy of plant biotechnology: 1942–2000. Madison, WI: University of Wisconsin Press.

Lacy, W. 2001. Generation and commercialization of knowledge: Trends, implications, and models for public and private agricultural Research and education. In: Wolf, S. and Zilberman, D. (Eds), *Knowledge Generation and Technical Change: Institutional Innovation in Agriculture*, pp. 27–54. Boston, MA: Kluwer Academic Publishers.

Le Buanec, B. and Ricroch, A. 2014. Intellectual property protection of plant innovation. In: Ricroch, A., Chopra, S. and Fleischer, S. J. (Eds), *Plant Biotechnology: Experience and Future Prospects*, pp. 59–73. New York, NY: Springer.

Lei, Z., Juneja, R. and Wright, B. D. 2009. Patents versus patenting: Implications of intellectual property protection for biological research. *Nature Biotechnology* 27(1): 36–40.

Lessig, L. 2004. The creative commons. *Montana Law Review* 65(1): 1–13.

Levins, R. A. and Cochrane, W. W. 1996. The treadmill revisited. *Land Economics* 72(4): 550–553.

Link, A. N. and Siegel, D. S. 2007. *Innovation, Entrepreneurship, and Technological Change*. New York: Oxford University Press.

Liptak, A. 2013. Justice, 9–0, bar patenting human genes. *New York Times*. Available at: https://www.nytimes.com/2013/06/14/us/supreme-court-rules-human-genes-may-not-bepatented.html (Accessed 13 August 2022).

Marks, J. H. 2014. Toward a systemic ethics of public-private partnerships related to food and health. *Kennedy Institute of Ethics Journal* 24(3): 267–299.

Mazzucato, M. 2013. *The Entrepreneurial State: Debunking Public vs. Private Sector Myths*. London, UK: Anthem Press.

Mirowski, P. 2011. *Science Mart: Privatizing American Science*. Cambridge, MA: Harvard University Press.

Merton, R. K. 1973. *The Sociology of Science: Theoretical and Empirical Investigations*. Chicago: University of Chicago Press.

National Academies of Science 2016. *Genetically Engineered Crops: Experiences and Prospects*. Washington, DC: The National Academies Press.

OECD (Organization for Economic Co-operation and Development) 1996. *The Knowledge Based Economy*. Paris, France: OECD Publishing. Available at: https://www.oecd.org/officialdocuments/publicdisplaydocumentpdf/?cote=OCDE/GD%2896%29102&docLanguage=En#:~:text=The%20OECD%20economies%20are%20increasingly,and%2 0learning%20in%20economic%20performance (Accessed 8 August 2022).

Pallottini, L., Garcia, E., Kami, J., Barcaccia, G. and Gepts, P. S. 2004. The genetic anatomy of a patented yellow bean. *Crop Science* 44(3): 968–977.

Pechlaner, G. 2012. *Corporate Crops: Biotechnology, Agriculture, and the Struggle for Control*. Austin, TX: University of Texas Press.

Piesse, J. and Thirtle, C. 2010. Agricultural R&D, technology, and productivity. *Philosophical Transactions of the Royal Society. Series B, Biological Sciences* 365(1554): 3035–3047.

Powell, W. W. and Owen-Smith, J. 1998. Universities and the market for intellectual property in the life sciences. *Journal of Policy Analysis and Management* 17(2): 253–277.

Samuelson, P. A. 1954. The pure theory of public expenditure. *The Review of Economics and Statistics* 36(4): 387–389.

Samuelson, P. A. 1955. Diagrammatic exposition of a theory of public expenditure. *The Review of Economics and Statistics* 37(4): 350–356.

Sandel, M. J. 2012. *What Money Can't Buy: The Moral Limits of Markets*. New York, NY: Farrar, Straus, and Giroux.

Sandel, M. J. 2021. *The Tyranny of Merit: Can We Find the Common Good?* New York, NY: Farrar, Straus, and Giroux.

Schenkelaars, P., Vriend, H. and Kalaitzandonakes, N. 2011. *Drivers of Consolidation in the Seed Industry and Its Consequences for Innovation*. Italy: COGEM.

Sease, E. J. 2007. History and trends in agricultural biotechnology patent law from a litigator's perspective. In: Kesan, J. P. (Ed.), *Agricultural Biotechnology and Intellectual Property: Seeds of Change*, pp. 38–44. Cambridge, MA: CAB International.

Slaughter, S. and Rhoades, G. 2004. *Academic Capitalism and the New Economy: Markets, State, and Higher Education*. Baltimore, MD: Johns Hopkins University Press.

Somers, M. R. and Block, F. 2005. From poverty to perversity: Ideas, markets, and institutions over 200 years of welfare debate. *American Sociological Review* 70(2): 260–287.

Stix, G. 2006. Owning the stuff of life. *Scientific American* 294(2): 76–83.

Strauss, D. M. 2009. The application of TRIPS to GMOs: International intellectual property rights and biotechnology. *Stanford Journal of International Law* 45(2): 287-320.

Walzer, M. 1983. *Spheres of Justice: A Defense of Pluralism and Equality*. New York, NY: Basic Books.

Wright, B. D. 2012. Grand missions of agricultural innovation. *Research Policy* 41(10)(17: 1716–1728.

Chapter 13

Ethics of agricultural research

N. Yasemin Yalım, Ankara University, Turkey; Burçin Çokuysal, Ege University, Turkey; and Cemal Taluğ, Ankara University, Turkey

1 Introduction

The transition to agriculture and settled life is one of the most important cornerstones in human history and can be considered the starting point for agricultural research. It can be said that the first agricultural research was empirical studies made by producers and based on observation. In fact, until the industrial revolution, it was always farmers who created and shared knowledge in agriculture. In addition to the cultivation and domestication of almost all plant and animal species grown today, all plant and animal breeding work before the industrial revolution, the use of animals as traction power and the development of new methods in irrigation are always the work of farmers. It is known that most of the studies in ancient astronomy, mathematics, and physics were carried out for the needs of agriculture; agricultural production is one of the driving forces for all basic sciences.

This chapter has three basic frameworks. First, it is essential to determine the scope and boundaries of the term 'agricultural research'. Second, we define the relationship between agricultural research and the concept of ethics to reveal the term that most accurately expresses the ethical dimension of agricultural

http://dx.doi.org/10.19103/AS.2023.0125.14

research. In this context, we argue that there is a distinction between terms that can be used interchangeably within the same context, such as agricultural research ethics, ethics in agricultural research, ethics of agricultural research, and agricultural research and ethics. The third point is to reveal the ethical subject areas and dilemmas that may be in question in agricultural research, based on the first two frameworks, and to make a general evaluation of them.

Agriculture is an applied science that deals with the production of plant and animal products, increasing agricultural productivity, the quality of the products, proper harvesting and preserving of these products in appropriate conditions, and processing and marketing them. With this structure, it is similar to other applied sciences such as medicine and engineering. Like them, agriculture also uses the knowledge produced by the basic sciences in which the primary concern is to understand and explain the universe and the world to make changes more favorable for humanity. Therefore, it is necessary to distinguish between agricultural research and basic science research, whose findings are also used in these studies. For example, while deciphering and compiling the gene sequence of a seed would fall under basic scientific research, the comparison of seeds with different gene sequences in terms of productivity, resilience, etc., falls within the scope of agricultural research. We think that this leads to an important distinction in terms of ethical evaluation and we build the following text on this assumption.

In addition to this distinction between research in basic sciences and agricultural research, we also consider the relationship of agricultural research with economics, psychology, sociology, anthropology, archaeology, law, and communication. We think agricultural research is a multidisciplinary and interdisciplinary field that includes research on agricultural production, agriculture-nature-human relationship, and other subjects related to social sciences.

Agricultural ethics is a recently developing discipline of applied ethics. It can be said that among all the subjects of agricultural ethics, the ethics of agricultural research is the least studied area. There is no consensus on what the subject includes. Our perspective is Ethics in Agricultural Research should be understood as the testing of ethical approaches to agricultural research. Study of all aspects of agriculture and the Ethics of Agricultural Research should be used to define ethical issues that arise from agricultural practice and research. We predict the study of agricultural ethics will become increasingly important. We have used 'ethics of agricultural research' in this chapter to maintain the order of the book. We will open the issue for discussion after further reflection on the terms.

Scientific ethics includes responsible research, scientific integrity, plagiarism, duplication, and unethical authorship. These issues are very important, but we reserve limited space for them in this section, considering that the specific problems of ethics of agricultural research are very diverse and comprehensive.

We thought that more detailed information about ethical issues which are the common concern of all scientific research can be found in any research ethics book. In the following paragraphs, we will only provide a brief summary of general research and publishing ethical issues. We will briefly mention these general problems while dealing with the specific ethical problems of agricultural research whenever necessary and with examples. The primary focus will be on the ethical problems that arise in the process of agricultural research (e.g. choosing the subject, specifying objectives, determining methods, and sharing and using the information produced).

We clustered the ethical issues in agricultural research under a few rubrics to make it easier for the reader to follow our discourse. We will discuss the relationship between agricultural research and nature, science and technology, the effects of the evolution of the agricultural production models on research, and the facts regarding the governance and financing of agricultural research. Considering the wide scope of ethics of agricultural research, which actually deserves to be a book on its own or even a series of books, this chapter can only be a summary.

2 Agricultural research as a subset of scientific research

Ethics of agricultural research involves the application of fundamental ethical principles related to scientific research. These include rejection of scientific misconduct, such as fabrication of data and plagiarism and undeserved authorship. The core value of all scientific research is trust based on honesty, and respect for human dignity. Researchers should not doubt the accuracy of research results published by other colleagues. Society should trust that research outputs reflect scientists' efforts to understand the world in good faith and without bias. Maintaining trust is only possible if the scientific community internalizes research ethics and takes it as a guide. The researcher has a responsibility to the scientific community, society, and future generations.

In accordance with this, there is a duty to protect the rights of people in the study and their autonomy and privacy. It is the researchers' ethical responsibility to take precautions to avoid every foreseeable harm. Ethically sound research should also include ways to compensate for the harm that may occur despite the measures taken.

Researchers have a responsibility toward science and the scientific community as mentioned above. They have to abstain from scientific misconduct which violates the codes of conduct and ethical behavior in scientific research. In a world of 'publish or perish', it takes a strong conscience and a solid moral compass. In agricultural research, as in many fields of science, exact reproduction of the results is often difficult if not impossible. All the investigations are dependent on the documents provided by the researchers. It is not possible to ensure scientific integrity in agricultural research without researchers' full cooperation. The only way to fight against scientific misconduct is by educating researchers and applying severe punishments in case of violating the codes of conduct with regard to scientific studies without tolerance (COPE, 2022).

A brief definition of the types of scientific misconduct is:

Fabrication - Reporting the results that were not properly obtained or the results of a study that was never conducted.

Falsification - Manipulating research materials, equipment, or processes or changing or omitting data or results such that the research is not accurately represented in the record. One form of falsification is pruning, manipulating the data by including only the results which support the hypothesis. Another is normalization, manipulating the data in order to achieve statistically more accurate results.

Plagiarism - Appropriation of another person's ideas, processes, results, or words without giving appropriate credit. There are various actions under this rubric such as copying a study completely or partially without referring to the original study, using various parts of different studies without properly referring to them, or copying figures from published articles.

Self-plagiarism is a different and complex version of plagiarism which occurs in several forms such as publishing a number of articles on the same subject, using the same set of data or using the same chunks of text again and again without referring to the previous ones.

Duplication refers to publishing the same intellectual material more than once. While this is not considered stealing, it is not an ethical attitude. To publish the same article in different journals with different titles or in different languages are versions of duplication. There is debate as to whether it is a form of self-plagiarism, but it should be acknowledged that it causes harms by creating article pollution in scientific literature.

Salami slicing of data is another form of scientific misconduct where a researcher slices data horizontally (in smaller numbers) or vertically (including fewer parameters) to produce multiple articles. The difference between these repetitive manuscripts is not adequate for evaluating them as separate and original articles.

There are numerous ethical issues about authorship, two of which are ghostwriting and guest writing. Ghostwriting refers to a phenomenon where the manuscript is written by someone other than the named author(s). Generally, the real author is hired by the sponsor of the research to formulate and report the research results in the best interests of the sponsoring institution or firm.

Guest writing is when the list of authors includes those who have not made substantial contributions to the research. Guest writers are either well-known academics who are listed as authors to ensure that the manuscript is published in a prestigious journal, or people who use their authority to get their names on the list, although they do not contribute to the work (Zietman, 2013; Ramesh, 2008).

3 The ethics of agricultural research: definition and scope

Agricultural research deals with the questions and problems that arise with the agricultural activities of people in nature or in non-natural production areas (vertical agriculture, soilless agriculture, aquaculture, etc.) to produce agricultural products and to further develop agricultural initiatives. Due to social, political, economic, and cultural effects, the scope, content, subjects, and methods of agricultural research have changed significantly and have become increasingly complex. The boundaries between basic sciences and agriculture have become more ambiguous in recent years due to the developments in genetics and biotechnology. Due to questions that have arisen in society, agricultural research has or should become multidisciplinary, including psychology, sociology, economics, public administration, city and regional planning, law, philosophy, and ethics in addition to science and research fields such as biology, chemistry, medicine, mathematics, and engineering.

For example, Luca et al. (2015) have presented the conceptual and methodological framework of an Italian research project titled 'MIMeSMAS', namely Multidisciplinary and Innovative Methodologies for Sustainable Management in Agricultural Systems. The project seeks to define an integrated approach for the assessment of environmental, economic, and social sustainability of innovative agricultural practices in Mediterranean regions, with a multidisciplinary, multi-methodological, systemic, and participatory approach.

Trnka et al. (2018) put forward the priority questions in drought research, which is one of the important consequences of the global climate crisis, from a multidisciplinary perspective. Monitoring of drought with remote sensing techniques, its effects, climatology, adaptation, planning and political aspects, as well as drought-society interactions were also examined. Researching a holistic and multidisciplinary approach to rehabilitation of degraded rangelands, Turnbull et al. (2021) reported that the products obtained from agricultural biotechnology have become one of the largest agricultural trade products globally, providing people with clothing, feed for animals, and fuel for our environmentally friendly cars; therefore, they are the main criteria in risk assessment with multidisciplinary studies.

Diptaningsari and Martono (2021) reported that agricultural research ethics is linked to personal, research, and social ethics, and this includes all activities before and after research. They stated that the advancement of agricultural biotechnology resulting in genetically engineered products has influenced studies on agricultural research ethics. They emphasized the importance of ethics in the use and development of genetic engineering in agriculture in order to avoid the negative effects of these products such as increasing environmental hazards, increasing human health deterioration, and unfair economic competition.

Kuiken et al. (2021) report that research on the development of CRISPR-based technologies has revolutionized genome manipulation while enabling the broad scientific community to redefine research and development for biotechnology products spanning food, agriculture, and medicine.

These changes in agricultural research should have led to the consideration of ethical problems and evaluation methods in this field. However, what appears to have happened is that consideration of agriculture's ethical problems has occurred outside the agricultural disciplines whereas it should have occurred outside and inside. It is inevitable there will be differences between past and present ethical evaluations of agriculture's ethical problems. These differences need to be taken into account.

Grimm (2011) categorized ethical studies in agricultural production and research as (a) the approach that tends to apply ethical theories and related norms to the field of agriculture and food, and (b) the approach that focuses on choices at all stages of research, starting from the research topic. It seems appropriate to refer to the first approach as ethics in agricultural research and the second approach as Agricultural Research Ethics. In this section, we take the second approach. In order to improve the discourse in the scope of agricultural research ethics, it is imperative that the subject be discussed in agriculture, especially among the practicing agriculturalists as well. This discussion is still limited if not missing and should be encouraged with studies such as this book.

Thompson and Kaplan (2014) defined agricultural ethics as a subset of applied ethics that addresses ethical issues and questions that arise in connection with agricultural activities and the production and distribution of food products. In the context of this definition, the scope of agricultural research is limited to the present technologies and related fields used to realize agricultural production. However, agricultural research needs to be conceptualized in a much broader framework because of its far-reaching connections to every aspect of life, including agricultural production technologies and bio-industrial animal husbandry, protection of natural resources and environmental problems arising from agricultural production, hunger, food safety and security, social policies, and legal regulations developed at all stages from farm to fork.

It should be taken into account that agricultural research ethics has changed in terms of its scope and values. It is a fairly new academic field and is possible to observe these changes. For example, the limited number of ethical evaluation examples made around the eighteenth century considered the resources of the world to be infinite and regarded the effects of man on nature as negligible because of nature's self-repairing power. The effect of humanity on other living things was almost never mentioned. Man's intervention and conquest of nature was considered as a right. However, there have been significant changes in these assumptions, especially since the second quarter of the twentieth century. The power and influence of man came to be considered as a threat to nature. Sustainability, environmental effects now dominate ethical evaluation. The factors that triggered those are briefly mentioned below.

(a) Change of perspective: The understanding that both natural resources and nature's ability to repair itself are limited has led to questioning the intensive agricultural production paradigm based on the classical productivity approach. The awareness that increased productivity has irreversible effects, as well as economic gain, has led to the conclusion that a different perspective of values is needed both because of the environmental and health problems that emerge due to classical productivity understanding and because these problems do not recognize the borders of countries that artificially divide our planet. This intellectual change has required a wider perspective that protects the environment, natural resources, and biodiversity. Sustainable production, healthier food is now evaluated and justice, solidarity, and trust are mandatory.

(b) A holistic approach to the ecosystem: Due to environmental pollution, global warming, climate change, and deforestation, all caused by human activities, the world may become incapable of supporting not only human life but also other living things in the foreseeable future if immediate action is not taken. For this reason, it is imperative for humans to make smart choices for the future of both their own species and nature. While this ethical responsibility imposes the obligation to consider the ecosystem as a whole, it also requires agricultural research be founded on these criteria during ethical evaluation.

(c) Globalization of the economic system: Agriculture, the food systems, and agricultural research have been affected by globalization and the free market economy. The food chain was quite short in the past, the longest distance from farm to table was from village to nearby city. Nowadays, it has become the norm for the food grown in one part of the world to be processed in another country, packaged in a completely different country, and delivered to the table of people in another country.

The globalization of the economic system has resulted in increasing commodification, financialization of agriculture, and monopolization of agricultural inputs and information. These changes determine not only the content of agricultural research but also its fundamental values in ethical evaluation. Ethical problems, including food sovereignty, culturally appropriate food, land grabbing, and fair trade, must be considered.

(d) Industrial production: The replacement of traditional agricultural production models by systems that provide agricultural inputs from all over the world and bio-industrial production is shaping agricultural research. Producers are becoming dependent on transnational companies, high-input chemicals, and genetically modified seeds and varieties to produce high volumes of low-cost output. This changes the subject content of agricultural research, leads to the adoption of different channels and methods in terms of financing research, and causes the commodification of agricultural knowledge. New problems emerge: can living organisms be patented, who owns research information, are research funding systems ethically appropriate or some better than others, and how do we diagnose and deal with conflicts of interest?

(e) Increasing role and power of non-governmental organizations (NGOs): NGOs have recently turned their interest to agricultural research issues, in addition to their active work on agriculture, environment, and nature. It is an indication that social awareness and sensitivity are developing and humanity is becoming conscious of its ethical responsibility to nature. These organizations establish food communities in cities, aid local, organic agriculture, and develop agricultural research. They also work to prevent gender discrimination and child labor in agricultural production, protection of animal rights in industrial animal husbandry, food sovereignty, and food democracy. They undertake supportive, ethically compelling functions. They implement 'The Power of Common Sensitivities', perhaps as a counter-alternative to the concept of the 'Tragedy of the Commons (Hardin, 1968)'.

Another function of NGOs is to be involved in the process by questioning the ethical legitimacy of the purpose and method in agricultural research and evaluating what ethical problems the results may cause in terms of both ownership and affecting existing relationships. For instance, Thompson's article (1988) covers the ethical analysis of the research on the tomato picking machine and highlights the social problems that may arise as a result of the use of this machine. The ethical sensitivities of civil society can be important guides in the ethical evaluation of agricultural research.

Jack et al. (2022) studied MASIPAG, an NGO in the Philippines that promotes sustainable agriculture based on agroecology and revealed its impacts on the mainstream economic policies and the process, work, and organization of extending the material freedoms of farmers affected by the climate crisis. They examined the network of alternative organizations from the Global South, showing how agroecology offers an alternative, more sustainable, and equitable future for smallholder farmers.

Designed as a general approach to networked participatory research aimed at supporting agroecological intensification of small-scale farming in ten countries in Africa and the Andes region of South America, Richardson et al.'s study (2022) established the necessity of farmer research networks. It was concluded that establishing links between complementary individuals and organizations could potentially improve the quality of rural life through multiple roles and tackling specific problems in agricultural production.

Mollah et al. (2021), in their study examining the role of NGOs in local agriculture in Bangladesh, revealed that regardless of the fact that NGOs are providing agricultural support to farmers, government agricultural extension institutions' active involvement for training farmers on how to improve agricultural production as well as raise awareness among the farmers to identify suitable location for crops beforehand is necessary for the improvement of the agricultural sector.

Challa et al. (2021) stated that 80% of economically active women in India are in the agricultural sector. They emphasized the importance of empowering women in agricultural production to meet the demand of the current and future populations, as young people migrate to urban areas. They stated that NGOs play a very important role in ensuring that women farmers in countries have equal access to productive resources with men.

(f) The social dimension of agriculture: The scope of agricultural research is determined by the classical production and profit understanding. This has reduced agriculture, which is one of the most basic activities of human beings, to the value of productivity and has made its socio-cultural and ethical dimensions invisible.

There are some issues that need to be discussed in regard to whether they are in the scope of agricultural research, such as the shrinking of villages as people leave and farms expand, which has become an increasingly important problem since the second half of the twentieth century; dehumanization and the ruthlessness of agriculture; hunger as the equivalent of poverty in agriculture and food system; education, health, and cultural services that rural areas are deprived of; gender inequality; and multidimensional inequalities in income distribution. The authors of this section argue that these issues, which are in the intersection of agriculture with social science fields, should also be included in the concept of contemporary agricultural research, and that agricultural research ethics should also cover these issues. We believe that when we exclude social fields from agricultural research, it will become ethically questionable as a concept.

4 Ethical issues in agricultural research

When today's agricultural production models are examined, two basic production methods are observed globally. The first one is a capital-intensive and monoculture-based 'industrial production model', which uses chemicals and seeds marketed by the world's transnational companies, technology-based tools, and devices instead of human and animal labor, including gene technologies, satellite data, computer-assisted irrigation systems, and artificial intelligence decision-making systems. This causes widespread pollution in terrestrial, aquatic, and atmospheric ecosystems due to intensive use of fossil fuel and its lack of sustainability. It increases off-farm dependency and loss of medium and small farms due to lack of financing and increasing debt. Industrial agriculture is not sustainable due to its burden on nature, social destruction, and the drive 'get big or get out'.

The second model is called agroecology and defined as a science and a movement in addition to being a practice. Being a contemporary way to practice agriculture, it includes conservation agriculture, permaculture, and organic agriculture and emphasizes the unity of plant and animal production, species and genetic diversity, reduced energy input, natural processes, and biological pest management. It is designed to maintain a functioning agroecology using biological, biophysical, ecological, social, cultural, economic, and political knowledge. It shortens the cognitive, social, physical, etc. distance created between producers and consumers in the Western world and is designed to create an ecologically sustainable, socially equitable food system (Wezel et al., 2014). Current agriculture is shaped by these two production models.

Industrial agriculture predicts producer's income and world food supply will increase by increasing production. Agricultural research is designed to constantly increase production. Its problem areas include 'factory farms', mass production of a single product, the replacement of labor by machines, elimination of small-scale family farms, and being harmful to nature and other species. It dominates agricultural research and separates man from nature. Agricultural products brought from all over the world are sold in cities with intensive marketing strategies. The system divorces consumers from agricultural products and producers. Food has become a commodity similar to cars and computers. To achieve its production goals, agricultural research includes chemical inputs and dependence on international companies. To counter objections, the system emphasizes the necessity of achieving sustainability as a primary objective. Researchers basically aim to make the existing system sustainable by any paradigm change rather than making radical changes in the industrial model.

In contrast, agroecological research focuses on local food production, cooperative or community-supported agriculture with low external inputs and short food chains. The research is based on two basic values.

First is achieving effective and efficient food production using agroecological principles. The second one is reducing the negative effects of the industrial model on the environment and the planet by putting nature at the center. It emphasizes the differences between industrial production and agroecological production. Research on the social effects and producer-consumer solidarity is important in agroecological research. The ethical problems of agricultural research in the opposing production processes are briefly summarized.

5 Ethical problems related to the subjects of agricultural research

As we have emphasized, the main research subject of industrial agriculture is increasing productivity and reducing production costs. The facts that ethically justify these issues are hunger (a shame on humanity) and poverty which is the primary cause of hunger. Research on increasing productivity has become ethically questionable because of environmental problems, climate change, and decreases in biodiversity. Research in recent years has aimed to reveal new technologies that can replace environmentally harmful technologies and reduce consumption of natural resources (Korthals, 2017).

Agricultural research increasingly emphasizes elimination of stress and transfer of genes that resist stress factors. Studies on stress factors show positive effects on agricultural production. In their study, Sahil et al. (2011) stated that the expression and regulation of stress-sensitive genes are essential for plants to develop stress avoidance and/or tolerance mechanisms.

Shabbir et al. (2021) reported that sugar cane, which is one of the most important agricultural industry products contributing to the world sugar industry and biofuel production, can potentially be improved against environmental stress factors, especially with technologies such as molecular marker-assisted breeding, sugar cane transformation, and genome editing.

Yang et al. (2021), described in their research the main drought stress signals and signal transduction pathways in plants, and listed functional genes and regulatory genes related to drought stress.

Evaluation of the ethics of this research reveals that the main problem has been ignored and that the proposed solutions address only current problem. For example, most of the studies on genetically modified plants are on transgenic plants with heat and drought stress resistance genes. Elwafa and Shehzad (2021) investigated the effects of drought and heat stress, which negatively affect the growth, development, and yield of plants, at the gene level in wheat.

Yu and Zhang (2021), who investigated the effect of gene transfer on drought stress in transgenic wheat, reported that overexpression of TaWRKY46 in wheat resulted in increased drought stress tolerance.

Lehretz et al. (2020) aimed at co-expressing hexokinase 1 from *Arabidopsis thaliana* (AtHXK1) in guard cells and SELF-PRUNING 6A (SP6A) using the leaf/stem-specific StLS1 promoter in order to increase water use efficiency as well as tuberization under drought and heat stress in order to increase the abiotic stress resistance of potato plants. In their co-expressed research, they reported that gene expression for drought and heat stress tolerance provided a new strategy to improve abiotic stress tolerance in potatoes.

Şanlı and Gökçe (2021) examined the effect of miR160 by overexpression alone or under a combination of heat and drought stresses in potato cultivars and reported that physiological characteristics develop in transgenic plants overexpressing miR160a-5p under abiotic stress conditions.

Although it seems like a good solution to counter the effects of the climate crisis, they are actually solutions that save the day. Future studies should lead to changes in the practice of industrial agriculture and its contributions to climate change.

Research focusing on increasing agricultural production has prepared and organized soil conditions and fertilization programs for the best yield; new genetic studies are trying to make varieties suitable for the environment and existing soil conditions (Cho and Choi, 2009; Qui et al., 2021; Imran et al., 2021; Truong et al., 2021). Intensive research is expensive and requires technical knowledge. Without adequate discussion, it risks ignoring or avoiding possible ethical problems. For example, terminator seed technology is a new and patented method of preventing plants from producing viable seeds. This method has a serious negative impact on agricultural systems, biodiversity, and food security, especially in developing countries, and makes producers dependent on global seed companies. To increase this dependency, investment of seed companies in terminator seed technology research emerges as an important ethical problem area.

The ethical alternative to this approach seems to be a modified version of classical methods, which are reconsidered due to the risks of genetic modifications and the negative effects of intensive fertilization programs on the environment. For example, while the classical method focuses on cultivar selection depending on their 'water-use efficiency' for yield and quality, research on new approaches emphasizes 'effective use of water' in the selection of varieties under water-stress conditions rather than 'water-use efficiency' (Blum, 2009).

Most agricultural research focuses on plant nutrition to achieve maximum plant yield. It is clear this will have a significant effect on environmental pollution. On the other hand, relatively new integrated nutrient management and integrated soil fertility management consider the effectiveness of factors other than yield to produce more ethical solutions (FAO, 1995; Gruhn et al., 2000; Das et al., 2015). It is ethically more acceptable to prepare special research programs according to the region and its structural and cultural characteristics, instead of recommending standard agricultural research methods and monoculture agricultural practices globally. This can only be achieved through participatory approaches. We recommend development and implementation of on-farm trials with FSR/E[1] which includes farmer participation.

Sustainability is the prominent value in almost all of the research carried out by industrial agriculture (Hrustek, 2020; Wegren, 2021; Martos et al., 2021; Durant, 2021; Khan et al., 2021). Values such as justice, solidarity, and respect for the rights of future generations are usually not included in these studies. There are serious ethical problems, especially regarding the autonomy of researchers and the ownership of the information obtained.

Similarly, the joint study of the World Watch Institute and Nourishing Planet (2011) has shown that success can be achieved by focusing on planting frequency, seeding depth, changing irrigation systems and regimes, micro-dose fertilizer applications, integration of plant and animal production, management of stress conditions, and examples that applied in Africa. We recommend these studies instead of exclusive focus on genetic modification.

The focus of agricultural research on the development of industrial agriculture on a global scale also overshadows agricultural research that can produce local solutions, while creating a serious imbalance of power. Therefore, it is clear that a flexible approach that avoids taking a fixed, standard, and precise position in the selection of agricultural research topics can produce important production solutions and avoid the emergence of ethical dilemmas. It is important to choose agricultural research areas according to the socio-economic structure and the need of solutions specific to the region and resources of different producers.

The leading subject of agroecological research is to achieve the production level reached by industrial agriculture without contradicting the agroecological production principles, or at least to reveal the data proving

[1] FSR/E (Farming Systems Research and Extension) has focused on limited resource farmers. It has sought to develop a basic system for research and extension with the strong involvement of the farm families themselves in all stages of research and development with the objective of increasing the development of appropriate technology suited to a group of homogeneous farm.

that the increasing world population can be fed with this production method. In addition to conducting studies that support ecological sensitivities and acceptable production in terms of cost, agroecological researchers are the ones who conduct most of the social science studies in agricultural fields. Research questions such as revealing alternatives to technologies that threaten the sustainability of small farmers or family farming and facilitating access to production tools should be a focus of research. In this context, it should be noted that values of justice, solidarity, sharing, and respect for the rights of future generations are commonly included by this group.

It was announced at the Intergovernmental Panel on Climate Change that agricultural production is responsible for one-quarter of the continuous increase in greenhouse gas emissions. However, studies have shown that not all agricultural production systems have equivalent effects on global warming (Brenda et al., 2011; Kumar et al., 2021). While industrial agriculture and animal husbandry contribute significantly to global warming, they also represent the vast majority of greenhouse gas emissions from agriculture (Brenda et al., 2011; Sanders and Webber, 2014; Clark and Tilman, 2017; DEFRA, 2022). It is known that ecological-based methods of agricultural production used in medium and small-scale farms consume much less energy and emit fewer greenhouse gases than industrial agricultural production (Brenda et al., 2011; Altieri et al., 2012; Altieri et al., 2015; Dabkiene, 2017; Golasa et al., 2021; Aid, 2021). Agroecological management techniques have the potential to capture more greenhouse gases than industrial agriculture while producing less direct emissions. To reduce the effects of agricultural production on the global climate, agricultural research seems to be ethically correct and necessary for small-scale or local agricultural product management, livestock, pasture management, and local food systems (Brenda et al., 2011).

Clean and usable waters are scarce resources in the world. Since agriculture has a share of approximately 73% of global water use, it also has the greatest responsibility in terms of conscious water use and water protection. The need for agricultural research that leads to plant growth with low water footprints in suitable places ensures the widespread use of economical (pressurized) irrigation systems and support practices to prevent groundwater pollution.

6 Ethical issues related to the management and financing of agricultural research

Like all scientific research, agricultural research also requires adequate funding, advanced research infrastructure, and trained manpower. The scientific and social validity of agricultural research largely depends on the independence of the scientists and research institutions that carry out the studies. The fact that the results of the studies are respected by the scientific community also depends on the scientific honesty of the researchers and the fact that the research is carried out independently of external influences. For these reasons, agricultural research should be evaluated ethically. The support given to agricultural research and the institutions where the research is carried out should be clearly stated to avoid any doubts, especially the suspicion of a conflict of interest.

The first GMO product introduced to the market on a large scale was bovine growth hormone (rBGH) and is an example of the GMO controversy. BGH is a naturally occurring hormone in dairy cows; it is also known as bovine somatotropin. Monsanto chemical company first produced recombinant growth hormone. It is injected into dairy cows to increase milk production. The United States does not require labeling of rBST products. It has been banned in the European Union since 1990, illustrating the political and ethical controversy surrounding most genetically modified agricultural products. The FDA declared that it was not a threat to human health. The patented name of the rBGH hormone Posilac was marketed in the sales advertisements with the findings that the milk yield of the animals would be 30% higher with the regular injection of the drug. The harmful effects of the hormone, which causes 30% milk increase in cows, were revealed two years later. Producers reported that treated cows do not produce as long, do not have many lactations, and have serious hoof and udder infections. Moreover, cows had to be treated with the use of more antibiotics to eliminate these problems. In 1991, a researcher at the University of Vermont leaked information about severe health problems affecting rBGH-treated cows, including udder and hoof diseases, mastitis, and deformed births (Ferrara, 1998; Ewall, 2021). Thereupon, the manufacturer (Monsanto) promised the university more than half a million dollars for further research. The university declared its results were incorrect. The U.S. General Accounting Office (GAO) investigated the claims. During the investigation, the FDA stalled in providing the GAO with original Monsanto test data and the GAO was unable to obtain critical data from the University and Monsanto. The GAO terminated its investigation, concerned that Monsanto had had time to manipulate the questionable data and that any further investigation would be fruitless. In an effort to dissipate public concern, University of Vermont scientists finally released information showing rBGH's negative effect on cow health, years after the findings had been made. And years later, a university official announced that the negative effects of rBGH were hidden. Afterward, it was reported that rBGH causes prostate and breast cancer in humans as well as its known side effects, so milk containing rBGH was banned in the European Union in August 1999 (Engdahl, 2007).

The rapid acquisition of knowledge obtained through agricultural research has high potential to create economic added value and has changed the financing models of these studies. Powerful companies use their capital, technology, and political support to determine the direction of agricultural research. While the number and effectiveness of leading public institutions in the field of agricultural research are gradually decreasing, the influence of private sponsors such as fertilizer/pesticide/seed companies is increasing. States have transferred agricultural research to multinational companies within the framework of the rules of the global economy. Public research has been privatized, or public-private joint studies have become widespread in many countries in the last 30 years. States' remaining function is regulating the use of information obtained and the market dynamics related to use.

Since the Enlightenment, universities have been the main environments in which knowledge is produced. Because the economic value of knowledge has increased, the production of knowledge has passed from universities that are more collaborative, open, and free to the R&D laboratories of large companies. The information gained becomes trade secrets and sharing and use of knowledge is often limited to the boundaries set by patents and copyrights. Publicly supported research institutions cannot compete with this structure. In this competitive environment, universities make their own decisions depending on their institutional structure and traditional values. Most of the universities modified their research structures to meet with multinational companies' demands, to stay in the play as an effective actor. They even risked being criticized for working like the R&D department of multinational companies. Still, they were somewhat sidelined, especially because they could not keep up with the pace of market-oriented science. A fundamental ethical problem created is the level of knowledge attained and the phenomena that need explanation (in Thomas Kuhn's words (1970), 'anomalies') that become visible during research are known by a very limited group of people and can be overlooked or neglected if they are seen only as an economic benefit. Delays and errors can regularly occur.

It is much more difficult for agroecological research to obtain sufficient financial support for large-scale research. Much of the required research is associated with sustainability and resilience to climate change. The studies should often be horizontal - extended over years rather than quick and easy. The strength of research groups is enhanced by sharing information obtained from studies and allowing researchers to conduct their research more freely. Sharing is usually accomplished by publication in open-access scientific journals. That is not as common among industrial researchers. It is clear that the funding of research is a matter of scientific integrity and should be taken into account in determining the scientific and ethical value of agricultural research. Public support of agroecological research is necessary.

7 Ethical issues regarding the sharing of agricultural research results

Agriculture, the essential human activity, is the most widespread human interaction with the environment. Agricultural information has been shared throughout history. There are great differences in geography; agricultural production has often been based on farmers learning from each other. The free exchange of information among agricultural producers has assured adherence to basic ethical principles (Giddens, 2009).

The understandable profit goals of industrial agriculture have resulted in conflicts of large against small producers, corporate farms against family farming, and multinational companies against local production. It is worrying to see that the new scientific norm is to hide knowledge rather than sharing it. When knowledge is used to provide labor and financial support for production to create added value, one encounters value problems that even lead to the patenting of living organisms.

The use of patented living organisms in agricultural production creates a power imbalance for small producers who cannot reach these products. In addition, producers using patented transgenic plant species produce with less input than producers who do not. Moreover, producers using local varieties are exposed to more disease and pest effects because the varieties they use are not resistant to diseases and pests.

Examples of plants that may be patentable include transgenic plants with enhanced properties such as resistance to pests, drought, or herbicides (e.g. herbicide-resistant soybean seeds or drought-resistant rice) and transgenic plants with health beneficial properties (e.g. golden rice, which is rich in vitamin A and used as a fortified food in regions where a diet deficient in vitamin A is consumed, or a transgenic rice that accumulates less arsenic in the grain).

The predominance of a transgenic variety poses a serious threat to biodiversity. An unfair competition environment is created for small producers who do not have the opportunity to reach these seeds.

Another example of value problems in this context is the relationship between the research on the tomato harvesting machine and the social problems that this technology will cause to dismiss local workers (Thompson, 1988). The opportunities provided to support agriculture financially raise a problem of justice, as they are hardly used by family farmers due to the complex bureaucracy and lack of knowledge, even in developed countries such as the European Union, and they serve to make large companies even more wealthy. In developing countries the governments that cannot withstand the pressures of multinational companies prohibit farmers from producing with local seeds and deprive those who insist on doing so for ethical reasons from financial support can be mentioned as another example.

We suggest some multinational companies see the exploitation of labor and natural resources in developing countries as a necessity of the global economy increases the ethical problems.

Since the 1970s, the rise of global capitalism has revealed that those in developing countries can profit from lower labor and environmental and human rights standards. Research on the subject raised the question for Western multinational companies, 'How should a company behave when the standards followed in the host country are lower than the standards followed in its home country?' (Monshipouri et al., 2003).

In the 1960s and 1970s, companies in advanced economies began relocating their manufacturing facilities and contracting in developing countries to reduce labor costs (Ast, 2019).

Technological progress has allowed rapid growth by changing people's environments. Although this situation has some benefits, it also brings with it ethical problems related to the use of natural resources in countries with weak environmental protection programs (Asgary and Mitschow, 2002).

The Texaco company has been accused of causing irreversible damage to the Ecuadorian rainforest for using low standards. A similar case occurred at the Vaca Muerta shale gas field in southern Argentina. Protests arose against the activities of the oil company Chevron for using low environmental standards (Livingstone, 2016).

Another common international environmental problem is related to the storage of toxic waste. It is estimated that a significant portion of China's arable land is contaminated with lead, zinc, and other heavy metals exported from developed economies (Buckley, 2011).

This leads to pollution of soil and water, loss of biodiversity, extinction of some species, and deterioration of ecological balance. Each of these problems has started to be evaluated as a separate agricultural research ethical problem area.

The problems that arise in agricultural research are not only economic. A tragedy in the history of agricultural research was caused by the ideological use and distortion of results due to T. Lysenko's hegemony over Soviet agricultural research. He created pseudoscience (Joravsky, 1962; Borinskaya et al., 2019). Another example would be the 'unscientific data' and 'exaggerated claims' put forth by the Indian Council of Agricultural Research , including the case of the Sharbati Sonora wheat strain that drove Indian Agricultural Research Institute agronomist Vinod H. Shah to suicide (Hanlon, 1974; Shahare and Roberts, 2020).

First synthesized by Othmar Zeidler in 1874, the potential of DDT (dichloro diphenyl trichloroethane) as an effective insecticide did not receive serious attention until the studies of Paul Herman Muller in 1939, a Swiss chemist who worked for the Geigy company. It was first used in the 1940s to combat insect-borne diseases. It is widely used as an insecticide after WWII in agricultural production. It was among the first synthetic organic chemicals used in agriculture and in the green revolution. Its rapid effective success and its widespread use in the United States and other countries led to resistance of many insect species. Because of its lipid solubility, it passed to other creatures such as chicken, sheep, and cattle (and humans) in the food chain with straw, grass, and grains and to the creatures living in rivers, lakes, and seas. Humans inevitably received DDT from foodstuffs (Berg, 2009).

Silent Spring, published by Rachel Carson in 1962, changed the course of history. She discussed the ecology of the environment and the human body. The book meticulously revealed the environmental destruction caused by state-approved pesticides used in agriculture; it discussed effects on the environment and natural life and human (Carson, 2002). The destruction of agricultural and natural areas became the focus of public attention. Carson was also the pioneer of the idea of biosphere health in agricultural research with the artistic expression of scientific knowledge (Özdağ, 2011). She emphasized the dangers of one-dimensional agricultural research.

Another example is the study by Dr. Arpad Pusztai and his wife Dr. Suzan Bardocz at the government-sponsored Rowett Institute in Aberdeen, Scotland, where the most important scientific results regarding the future of the genetically modified seeds project were obtained. Dr. Pusztai was considered one of the leading names in the field of genetic modification of plants, with his 35 years of experience in 270 published biotechnology articles and data in two books. Prior to the large-scale sales of GMO soybeans prepared for the American and Argentine markets in 1995, the Scottish Agriculture Office signed a 3-year contract with the Rowett Institute with a budget of US$1.5 million. The Scottish Agriculture Office asked the Rowett Institute to produce a mandate on scientific testing methods to help the government prevent and regulate potential future risks. Toward the end of 1997, the results of laboratory trials of Pusztai (1998) reported (Ewen and Pusztai, 1999) signs of weakening of the immune system and heart and brain and lung problems of mice fed with GM potatoes (Pusztai, 2003). Dr. Pusztai said: 'We can safely consume these foods, completely healthy and harmless substances, but as scientists we would never approve of our citizens turning into guinea pigs. The subjects should only be in the laboratory'. He stated that if he had to choose, he certainly would not consume these foods without obtaining evidence of their safety. He and his wife were fired from the Rowett Institute. They were able to get their pension rights on the condition that they never speak to the press. Although some of his colleagues and the Rowett Institute declared that Dr. Pusztai's research was wrong and the press published his research as invalid, the investigations proved that Dr. Pusztai was right. Pusztai disclosed his data as he conducted an independent study, funded by the public, without the sponsorship of any commercial concern. However,

Rowett's contract was not renewed, as his data was confiscated, and a ban was imposed on the subject. Thus, the denial of Pusztai's academic freedom is not only an ethical problem but also the collective intolerance of his view, which opposes the dominant view (Engdahl, 2007).

It is important to note the academy, which announced Dr. Pusztai's work as flawed, did not make any effort to replace these studies or to eliminate the mentioned deficiencies. Scientific impartiality lags behind some political pressures and interests.

8 Ethical issues that arise during the conduct of agricultural research

Given the wide scope of agricultural research, it would be wrong not to predict that value problems arise during the conduct of these studies. The main ones are:

(a) *Data ownership:* The first ethical problems that may arise during agricultural research is sharing of the benefits arising from the ownership and use of the information obtained. Great progress has been made from seed planting to harvesting due to digital farm management and detailed risk analysis. Development of smart agriculture systems and use of autonomous vehicles have improved crop yields and quality. However, ethical issues remain: Who owns big data obtained? Who can access the data for what purposes? Who benefits if data become a salable commodity? Who benefits? There are inevitable questions about abuse of data and the injustice that may arise for the producers who do not have equal access to the services developed from the data because of cost, lack of information, etc. (Mark, 2019).

Companies that analyze big data of farmers and provide agricultural technology collect a lot of data from farmers. With the analysis of these data, effective recommendations are offered to the manufacturers. The basis for providing effective advice to manufacturers lies in the fact that large repositories of data are required. The larger the quantity, type, and range of farm datasets available, the more detailed, accurate, and comprehensive will be the recommendations to producers. It is clear that big data provides a power in favor of companies that analyze data and provide agricultural technology. This 'power' is the ability to perform an activity without necessarily subordinating the autonomy of another. With short- and long-term projections, these companies can buy inefficiently productive land at low prices and implement restrictive and economically aggressive policies until they are forced to sell their land to a farm or partner company at a very low price. There is also the possibility of identifying problematic farmland and using this information directly or through one of its partners/affiliates to lower the farmer's asking price.

Companies that analyze big data and provide agricultural technology use the data of producers, while preventing producers and other companies from accessing this data. For example, Monsanto prohibits farmers from accessing data received by farmers; John Deere prohibits farmers from tampering with their machinery as it is seen as an intellectual property infringement; Pioneer and Monsanto require farmers to be their own seed customers first in order to advise them. This also undermines the autonomy of producers when using their data.

Another ethical problem area is whether producers' data will be shared with a third party. Because of ownership of this data, manufacturers may be forced to use products they may not want to use. For example, for farmers using Pioneer's Field360™ technology, it is mandatory to use Pioneer seed. FieldScripts® is currently limited to Monsanto's DEKALB seed varieties. Monsanto has also limited the use of FieldScripts technology to farmers whose companies use machines created by Precision Planting.

It is also invaluable to investors and financial traders who trade in wheat, soybean, and corn sales and futures by analyzing data from producers. In a market where the slightest information advantage makes the difference between huge profits and even greater losses, companies that collect big data will have a ready customer base if they choose to sell their information or use it to speculate on their own.

Carbonell (2016) states that multinational companies that monitor every decision made by farmers with sensors on their tractors have gained a privileged position on more than one-third of the farmland in the United States. They collect large amounts of proprietary data that provide unique, precise estimates and insights to a farm operation. Carbonell stated this power asymmetry can only be balanced with public research and development studies.

Ethical problems may appear in well designed, properly conducted research (Lewis, 2020; Ryan, 2020; Mark, 2019; Zampati, 2019). It is legal that recommendations to increase production capacity will be shared only with producers who provide data to the system. But it does not seem fair. Therefore, one must ask if the existing legal framework is fair and ethically valid. Because agriculture is so important, scientific autonomy, confidentiality, data ownership, benefit-harm, and unfair advantage should be carefully examined. When the collection and processing of large numbers of large databases to create large datasets in conducted research is open source and publicly available, malicious use of this data can be prevented.

(b) *Animal rights:* The ethical principles that apply to research on animals in medicine, biology, and genetics are valid in agricultural research in the

> field of animal husbandry. Animals can suffer and must be protected. Animal research should be done with the least possible number of subjects and the lowest possible level of pain and distress.

Although the responsibilities of the human species toward animals from past to present have been discussed on the basis of belief or philosophy, it cannot be said that these thoughts have guided agricultural research ethically. Systematic regulations in terms of ethics increased in agricultural research in the nineteenth century based on cruelty to horses, and working animals, and experiments on animals. A decrease was observed in social concerns about animals in agricultural research toward the end of the nineteenth century. Because animal rights reached a larger audience, animal ethics became a socially sensitive issue in the second half of the twentieth century (Simonin and Gavinelli, 2019; Duncan, 2019).

The book 'Animal Liberation' by Singer (1990) was a turning point in agricultural research ethics. After Singer's book, the concepts of 'animal rights', 'animal welfare', and 'animal liberation' were defended by many and made the ethical dimension of agricultural research programs visible. Singer criticized the industrial system that saw animals as machines to convert low-priced commodities into high-priced proteins at less. In addition, he strongly accepted that agricultural research on animals should be evaluated ethically. Singer's concern about animals and animal rights has not been widely adopted by other areas of agricultural research – e.g. pesticides, GMOs, soil erosion, and water mining.

According to some ethical approaches, there is no difference between studies investigating the effectiveness of not feeding chickens for 10–15 days to accelerate their laying cycle or of methods to shed their molt in industrial poultry and Nazi experiments trying to reveal how long a person can live under certain pressure and under how many degrees cold in terms of ethics. According to others, the difference does not justify the studies in question (Tuvel, 2015; Sztybel, 2006).

The priority in animal experiments is to carry out the research procedure correctly. But pain or suffering of animals is pushed into the background. The consistency of the moral justifications put forward to justify this situation is a matter of considerable debate. An ethical problem is whether animal experiments can be justified morally even if they will ensure the protection of the human species. Pharmaceutical and agricultural pesticide manufacturers defend animal experiments as they protect humans. They also point out that regulatory agencies require data from animal experiments (oral toxicity, eye effects) to approve the usage of the substances. Instead of continuing cheap animal experiments, it is possible to protect humans and other species with new technologies, new laboratory methods using artificial intelligence, and computer simulations (Passini et al., 2018; Badyal et al., 2009; Doke and Dhawale, 2015).

Acknowledging animals as entitled to some human rights might make it impossible to use them for nutrition or as subjects of research that cannot be done on humans. Because they can feel pain and distress, treating them as an object, a good, or a commodity cannot be ethically justified. Different approaches and methods have been adopted in other countries. For example, Austria prohibited considering animals as objects or goods in 1988 and Germany in 1990. In 2003, Switzerland decreed prohibited classifying animals as objects. The Constitution of the European Union describes animals as sentient beings, not property (Cardosa et al., 2019; Broom, 2017; Simonin and Gavinelli, 2019; Duncan, 2019). As a result, animal life has moved beyond the individual preferences of those involved in agricultural production and research.

While studies to reduce the negative effects of industrial animal husbandry on the welfare of animals are ethically positive and necessary, the issue of injustice that may arise among the animals used in these research is a continuing ethical debate.

(c) Genetically modified organisms, vectors, advanced chemicals, and similar agents used in agricultural research can have negative effects on the environment either by dispersal or by invasion of much distant places in the world. For this reason, safety and compensation procedures are extremely important and necessary in agricultural research as they are in medical research. Much agricultural research is carried out on organic material, and in this context, they may have both predictable and unpredictable effects on earth.

9 Theoretical basis for the ethical analysis of agricultural research

Humans used to benefit from nature by hunting and gathering when natural forces shaped nature without human intervention. Later, agricultural production turned into an activity of recreating nature with a socio-economic concern. Agriculture has never been just a business. It never will be and never should be. Agriculture is an essential human activity. It is value-laden and requires a comprehensive ethical justification.

Through the transition to settled life, people produced different plant and animal races from the animal species they lived with and the plants they grew for thousands of years. In the beginning, most people were agricultural. Later, after the industrial revolution (1760 to early 1800), producers and agricultural researchers who focused on production techniques and the problem of feeding the increasing population were separated. In the last two centuries, different methods have been proposed to bring producers and researchers together.

The basic problem of the ethical evaluation of agricultural production and research is the lack of dialog between producers and consumers that will lead to a sustainable continuation of agriculture. There are many publications dealing with agricultural research; there are few addressing agriculture's ethical foundations. This suggests that people who think about the ethical dimension of agricultural research take the legitimacy of agricultural activity for granted (Zimdahl, 2012).

Gremmen (2021) points out that a fully biocentric and ecocentric position makes agricultural production and research philosophically impossible. He emphasizes that the critique of the anthropocentric approach is inevitable for welfare and contract ethics. According to him, while eco-ethics, welfare ethics, and rights ethics integrate humans with nature and view all individual organisms as equal in terms of welfare and rights, they become an inappropriate moral compass that may make agriculture impossible. He shows that social contract ethics could not provide sufficient grounds for anthropocentric orientation. He proposes 'care ethics' as the most comprehensive and appropriate ethical position that makes agricultural production and food systems and related research ethically possible. Considering that ethics is one of the existential potentials of man and a product of the human mind, we adopt this approach that may face criticism of being anthropocentric in its essence. We propose a form of ethical evaluation built on this ethical perspective. We think an approach that excludes humans or equates them with all other forms of life in the world cannot ethically justify agriculture.

In general, care ethics is an ethical approach that emerged in the context of medicine with 'care' as its starting point. It focuses on acting with care. For health, it focuses on the values, priorities, vulnerabilities of the parties involved in a therapeutic relationship - especially the patient. The way these are expressed support equality in the relationship and solidarity in terms of being human. In another article, Gremmen (2021) argues that agriculture is an interface between humans and nature and that this perspective can be applied to our relationship with plants, animals, soil, in short, all areas of agricultural production. He emphasizes the necessity of evaluating the ethical dimension of agricultural production and therefore of research in this field, through the relationship between man and nature.

Careful reasoning results in the responsibility to protect nature and other living things, not a right to exploit them. In this sense, the wellbeing of plants, animals, soil, water, the entire planet is a responsibility that we have because we are rational beings. That is, because we are human. As Gremmen emphasizes, it should be emphasized that living things, most of which are domesticated and therefore cannot survive without the care of their producers, are entrusted to the farmers by the society on behalf of humanity. His approach develops a care framework for agricultural production and research. It is based on four key elements.

(a) attention (careful observation of the entire ecosystem);
(b) responsibility (awareness that what is we observed may compel care functions);
(c) competence (performing caring functions in accordance with professional standards); and
(d) responsiveness (being alert to possibilities of neglect, abuse, or inadequacy and acting accordingly to remedy deficiencies) (Gremmen, 2021).

It is not possible to justify agricultural production and agricultural research in an ethical context without using the basic concepts of ethics: rights, justice, honesty, and benefit. It is not possible to make agricultural production and research sustainable within the framework of a human–nature relationship without abandoning the prejudice that people are the owners of nature and have the right to use it as they wish in an ecosystem where natural resources are rapidly consumed by exploitation. It is clear that agriculture must have a moral philosophical foundation.

The survival of the human species is only possible through agricultural practice in nature. Agricultural studies in physiology, biochemistry, genetic engineering, agriculture, physics and chemistry affect production, cultivation, plant nutrition, irrigation, plant protection, mechanization, seed growing, soil structure, and healthy food production. All are affected by and influence the relationship between man and nature. It is becoming increasingly clear we should go beyond the idea that agricultural production's goal is only to meet human needs for nutrition. It should include the value of the natural, the responsibility of man toward nature, our relations with other species, and the effects of agriculture on the socio-cultural and economic life of societies (Korthals, 2016).

The agricultural production period focused on high efficiency and profitability in which the rules of classical development economics dominated. Agricultural research, as one of the pillars of industrial growth, had similar goals. Agricultural research on integrated rural development strategies and reduction of poverty and inequality has been abandoned due to the low profitability and high input costs. As a result, agriculture is not regarded as a development tool (De Janvry, 2010).

It has become necessary to change the agricultural production paradigm to reverse the environmental problems created by the intensive production paradigm. The sustainable agricultural production paradigm has been designed to protect natural resources and the environment and observe sustainable production practices essential to obtain high yield and quality products. This change has transformed the subject content of agricultural research (Welch and Graham, 1999; Byerlee et al., 2010). Instead of profit based on maximum yield, agricultural studies have begun to take into consideration economic factors such as the preference for optimum yield sensitivity to nature, social factors including increased living standards, equal opportunities in villages and cities, and awareness of the importance of protecting nature (Porter and Kramer, 2007).

Ethical evaluations have included environmental effects on use of natural resources, waste management, and pollution control into consideration during research and the necessity of sustainability.

In developed countries, which have been pioneers in agricultural research, the declines in the proportion of farmers, the share of agriculture in national income and the share of food expenditures in people's budgets have led to the gradual discrediting of agriculture and agricultural research. The effects of the climate crisis and the COVID-19 pandemic reinforced the importance of the value of agriculture. The food systems paradigm including efficient production, sustainable environment, and healthy food is becoming increasingly important (Welch and Graham, 1999; Kassam et al., 2010). It will all change the ethical approach to agriculture and agricultural research.

Evaluation of the four principles of organic agriculture revealed by IFOAM (2020) within the scope of agricultural research ethics seems promising for a sustainable and healthy food production paradigm. These principles are (a) Health should be taken into account. Soil, plant, animal, human, and planet health should be maintained and protected in a single and indivisible way for agricultural studies to be conducted; (b) Ecology. Agricultural research should take into account the need to maintain ecological systems based on living ecological systems and cycles and work together in harmony with them; (c) Justice. Agricultural research should build relationships that provide justice for the common environment and life of all species; (d) Care. Agricultural research should be based on the principle of prudent and responsible management to protect the health and wellbeing of future generations and protect the environment.

We are on the verge of a new era of science and technology in agricultural research. Research on artificial intelligence and smart farming systems will play an important role in shaping a future that may include research on extraterrestrial habitats. Discussing the ethical implications of these studies with all stakeholders is important as we decide what is right. It is necessary to use correct methods, common scientific and moral language, and appropriate forms of communication for ethical discussions to be effective and productive.

10 Conclusion

It is important to consider possible human and environmental risks and ethical problems that may arise in addition to the potential benefits of new agricultural research areas. It is no longer wise to justify all effects of agricultural research solely on providing more food for the growing human population. It is important to know the beneficiaries of research, consider the perspectives of all stakeholders in design and implementation of research, and discuss the relation of the research to nature, yield, and sustainability as part of an ethical evaluation of agricultural research.

Focusing on causes of problems rather than only on solutions to current agricultural and environmental problems will be more effective in the selection and funding of scientific research topics and will produce long-term sustainable solutions. For example, focusing on renewable energy research that will reduce the use of fossil fuels rather than trying to produce drought-resistant varieties caused by the global climate crisis seems more effective in slowing the effects of the global climate crisis.

Planning agricultural studies with a holistic perspective is essential and should be adopted from an ethical point of view. We must move beyond a human-centered perspective to agriculture and the environment as a whole with all living things.

It is clear that focusing only on food production in the agricultural industry, on exports or imports in agricultural research, cannot solve the problems of agricultural production today. It does not seem possible to produce sustainable and permanent solutions unless the studies are planned in a way that deals with the problems of stakeholders at every stage of agricultural production from field to plate in a wide framework and interdisciplinary structure, that can produce ethically grounded solutions that consider scientific, economic, social, and environmental value.

Agricultural research seems important in this context to contribute to the protection of biodiversity and local cultures, the adaptation of the ancient knowledge of producers to modern agricultural practices and similar purposes.

The industrialization of agricultural production, the usage and restructuring of big data by international companies, cartelization and commercial concerns inevitably create new ethical problem areas in agricultural production, and agricultural studies will take their share from these changes. These ethical issues should be addressed with an approach that recognizes the world is human's responsibility, agriculture is the farmer's responsibility, and that caring for both constitutes the core of ethical behavior.

Agricultural studies planned and carried out with this approach would contribute to a world where agriculture respects nature, labor, justice, and honesty that prevail at every stage of the food chain. Everyone will have sufficient, healthy food produced sustainably in food systems supported by science and ethics.

Who will determine research priorities will continue to be a problem for all scientific and agricultural research. However, the argument that humanity's problems must continue to dominate agricultural research, and nature and other species are important but secondary, does not see the ecosystem as a whole with all living and non-living beings. Hunger, violation of food safety and food security, negative conditions created by the global climate changes, extinction of species, environmental pollution, and the loss of agricultural land due to urban pressure are scientific and ethical problems. They all stem from the inability to see the ecosystem as a whole and to plan studies with this in mind.

Similarly, the exploitation of underdeveloped countries by developed countries, of local companies by global companies, of other species and nature by humans are ethically similar and not independent from each other. Seeing this injustice is an important ethical step in the planning of agricultural research. Exploitation of others and the environment is a moral choice and it is an important reason for the moral distress of those who produce scientific knowledge.

Studies carried out to end animal experiments in agricultural and other sciences stand out as an important ethical decision to protect the rights of all living things that cannot defend or claim their rights. We do not claim to prefer human life over animal life. We assert it is ethically proper to ensure that no living thing suffers unnecessarily. A choice with economic and social costs. Planning studies and educating society and researchers is important to the integrity of the ecosystem and the sustainability of the human species.

Undoubtedly, the aim of agricultural production is to meet the healthy food needs of humanity, provide the social, cultural, and economic welfare of the people, protect the environment and biodiversity, and achieve sustainability of the carrying capacity of the planet. Agricultural research is the basis of awareness and action on protection of these values. It indicates relationship that humans establish with nature.

The increasing commercialization of agricultural knowledge and the control of giant transnational corporations is a cause for concern. The public's share of financing of agricultural research and international cooperation should be expanded. Solidarity and fair sharing should also be among the main values of future research.

We think the award system has not been set up correctly in the majority of research institutions today. The number of publications should not be the primary criterion in the agricultural research award system. The pressure to publish or perish should be eliminated since it traps researchers in fashionable topics and research fields whose results can be obtained quickly. Award systems of agricultural research institutions should encourage and support long-term, creative, interdisciplinary, participatory research. The reward system should also consider the respect for nature and labor as valuable, in addition to increasing efficiency and production (Thompson and Stout, 2018).

Science and ethics should be placed jointly in the main axis of the agricultural research system in order to fulfill the responsibility of the agricultural researcher toward the scientific milieu, society, nature, and future generations in the 'good' and 'right' way.

Let us remember that 'values and ethics are no limit or obstacle to innovation and change; they are the gist of innovation and change. They represent the compass indicating what responsible, inclusive and sustainable ways of future-making are (European Group on Ethics in Science and New Technologies, 2021)'.

11 References

Aid, C. (2021). New reports show urgent need for agroecology revolution to cut emissions and boost productivity. Available at: https://reliefweb.int/report/world/new-report-shows-urgent-need-agroecology-revolution-cut-emissions-and-boost (accessed: 22 August 2021) (2018).

Altieri, M. A., Nicholls, C. and Funes, F. (2012). The scaling up of agroecology: Spreading the hope for food sovereignty and resilience. A contribution to discussions at Rio+20 on issues at the interface of hunger, agriculture, environment and social justice. *Sociedad Cientifica Latinoamericana de Agroecologia* 1-20.

Altieri, M. A., Nicholls, C. I., Henao, A. and Lana, M. A. (2015). Agroecology and the design of climate change-resilient farming systems. *Agronomy for Sustainable Development* 35(3): 869–890.

Asgary, N. and Mitschow, M. C. (2002). Toward a model for international business ethics. *Journal of Business Ethics* 36(3): 239–246.

Ast, F. (2019). The moral dilemmas of global business (Ed. George Yungchih Wang). *Globalization*. IntechOpen. doi: 10.5772/intechopen.74833. Available at: https: // www.intechopen.com/chapters/60620 (accessed: 22 August 2021).

Badyal, D. K., Modgill, V. and Kaur, J. (2009). Computer simulation models are implementable as replacements for animal experiments. *Alternatives to Laboratory Animals* 37(2): 191–195.

Berg, van den, H. (2009). Global status of DDT and its alternatives for use in vector control to prevent disease. *Environmental Health Perspectives* 117(11): 1656–1663.

Blum, A. (2009). Effective use of water (EUW) and not water-use efficiency is the target of crop yield improvement under drought stress. *Field Crops Research* 112(2–3): 119–123.

Borinskaya, S. A., Ermolaev, A. I. and Kolchinsky, E. I. (2019). Lysenkoism against genetics: The meeting of the Lenin All-Union Academy of Agricultural Sciences of August 1948: Its background, causes, and aftermath. *Genetics* 212(1): 1–12.

Brenda, B. L., Chappell, M. J., Vandermeer, J., Smith, G., Quintero, E., Bezner-Kerr, R., Griffith, D. M., Ketcham, S., Latta, S. C., McMichael, P., McGuire, K. L., Nigh, R., Rocheleau, D., Soluri, J. and Perfecto, I. (2011). Effects of industrial agriculture on climate change and the mitigation potential of small-scale agro-ecological farms. *CAB Reviews* 2011: 6 (020).

Broom, D. M. (2017). *Animal welfare in European Union*. Directorate General for International Policies. Brussels: European Union.

Buckley, C. (2011). Heavy metals pollute a tenth of China's farmland-report. *Reuters*. Available at: http://www.reuters.com/article/2011/11/07/us-china-pollution-agriculture-idUSTRE7A60DO20111107.

Byerlee, D., Janvry , A. and Sadoulet, E. (2010). Agriculture for development: Toward a new paradigm. *Annual Review of Resource Economics* 1: 15-31.

Carbonell, I. M. (2016). The ethics of big data in big agriculture. *Internet Policy Review* 5(1): 1-13.

Cardosa, D. S., Faraca, C. B., Sausa de, L. and Pereira, G. G. (2019). History and evolution of the European legislation on welfare and protection of companion animals. *Journal of Veterinary Behavior* 19: 64–68.

Carson, R. (2002). *Silent Spring*. Boston: Mariner Books Hougton Miffin Harcourt.

Challa, N. L., Jagadeeswari, B., Rakesh, K. and Naik, R. (2021). Intervention of NGOs and ITCs in mainstreaming Indian farm women in agriculture. *Current Journal of Applied Science and Technology* 40(14): 28–34.

Cho, E. K. and Choi, Y. J. (2009). A nuclear-localized HSP70 confers thermoprotective activity and drought stress tolerance on plants. *Biotechnology Letters* 31(4): 597–606.

Clark, M. and Tilman, D. (2017). Comparative analysis of environmental impacts of agricultural production systems, agricultural input efficiency, and food choice. *Environmental Research Letters* 12(6): 1-11.

Committee on Publication Ethics (COPE). Code of conduct and best practice guidelines for journal editors. Available at: https://publicationethics.org/files/Code_of_conduct_for_journal_editors_Mar11.pdf (accessed: 02 April 2022).

Dabkiene, V. (2017). A comparative analysis of on-farm greenhouse gas emissions from family farms in Lithuania. *Research for Rural Development* 2: 225–232.

Das, D., Dwivedi, B. S. and Meena, M. C. (2015). Integrated nutrient management for improving soil health, and crop productivity. *Indian Journal of Fertilisers* 11(4): 64–83.

Department for Environment Food and Rural Affairs (DEFRA). Agricultural statistics and climate change. Available at: https://assets.publishing.service.gov.uk/government/uploads/system/uploads/attachment_data/file/835762/agriclimate-9edition-02oct19.pdf (accessed: 2 April 2022).

De Janvry, A. (2010). Agriculture for development: New paradigm and options for success. *Agricultural Economics* 41: 17-36.

Diptaningsari, D. and Martono, E. (2021). Considering ethics in agricultural experimental research with special mention to agricultural biotechnology research. *Jurnal Perlindungan Tanaman Indonesia* 25(1): 10–20.

Doke, S. K. and Dhawale, S. C. (2015). Alternatives to animal testing: A review. *Saudi Pharmaceutical Journal* 23(3): 223–229.

Duncan, I. J. H. (2019). Animal welfare: A brief history. In: Hild, S. and Schweitzer, L. (Eds), *Animal Welfare: From Science to Law*, 13-20.

Durant, J. L. (2021). Commoning the bloom? Rethinking bee forage management in industrial agriculture. *Elementa: Science of the Anthropocene* 9: 1.

Elwafa, S. F. and Shehzad, T. (2021). Genetic diversity, GWAS and prediction for drought and terminal heat stress tolerance in bread wheat (*Triticum aestivum L.*). *Genetic Resources and Crop Evolution* 68(2): 711-728.

Engdahl, F. W. (2007). *Seeds of Destruction: The Hidden Agenda of Genetic Manipulation*. Quebec, Canada: Global Research Publishing.

European group on ethics in science and new technologies (2021). *Values for the Future: The Role of Ethics in European and Global Governance*. Brussels: European Commission, May 2021 (4).

Ewall, M. (2021). Bovine growth hormone: Milk does nobody good. *Web Resources for Environmental Justice Activist*. Available at: https://www.ejnet.org/bgh/nogood.html (accessed: 22 August 2021).

Ewen, S. W. and Pusztai, A. (1999). Effect of diets containing genetically modified potatoes expressing Galanthus nivalis lectin on rat small intestine. *Lancet* 354(9187): 1353-1354.

FAO (1995). Integrated Plant Nutrition System, *FAO Fertilizer and Plant Nutrition Bulletin* No. 12. Rome, 426 pp.

Ferrara, J. (1998). Revolving doors: Monsanto and the regulators. *The Ecologist*, September/October 1998. Available at: http://www.albionmonitor.com/9904b/monsantofda.html (accessed: 22 August 2021).

Giddens, A. (2009). *Sociology, (Revised and Update Sutton, W.P)*, 6th edition. Polity Press.

Golasa, P., Wysokinski, M., Golasa, W. B., Gradziuk, P., Golonko, M., Gradziuk, B., Siedlecka, A. and Gromada, A. (2021). Sources of greenhouse gas emissions in agriculture, with particular emphasis on emissions from energy used. *Energies* 14(3784): 1-20.

Gremmen, B. (2021). An ethical compass for agriculture. In: Yalım, N. Y. and Evren, M. (Eds), *III. International Agriculture and Food Ethics Congress*. Ankara: The Agricultural and Food Ethics Association of Türkiye, 31-37.

Grimm, H. (2011). Ethical issues in agriculture, interdisciplinary and sustainability issues in food and agriculture. In: Christen, O., Squires, V., Lal, R. and Hudson, R. J. (Eds), *UNESCO, Encyclopedia of Life Sciences*. Paris: ELOSS Publishers.

Gruhn, P., Goletti, F. and Yudelman, M. (2000). Integrated nutrient management, soil fertility and sustainable agriculture: Current issues and future challenges. *IFRPI 2020 Vision Brief* 67: 1-2.

Hanlon, J. (1974). Food scientist published false data. *New Scientist* 64(922): 436-437. Available at: http://swamiscapers.blogspot.com/2007/03/analytical-report-top-food-scientist.html (accessed: 22 August 2021).

Hardin, G. (1968). The tragedy of the commons. *Science* 162(3859): 1243-1248.

Hrustek, L. (2020). Sustainability driven by agriculture through digital transformation. *Sustainability* 12(20): 1-17.

IFOAM (2020). *Principles of Organic Agriculture*. Bonn: IFOAM Organics International.

Imran, Q. M., Falak, N., Hussain, A., Mun, B. G. and Yun, B. W. (2021). Abiotic stress in plants; stress perception to molecular response and role of biotechnological tools in stress resistance. *Agronomy* 11(8): 1-20.

Jack, G., Plahe, J. and Wright, S. (2022). Development as freedom? Insights from farmer-led sustainable agriculture non-governmental organization in the Philippines. *Human Relations* 1-28.

Joravsky, D. (1962). The Lysenko affairs. *Scientific American* 207(5): 41-49.

Kassam, A., Friedrich, T. and Derpsch, R. (2010). Conservation agriculture in the 21 century: A paradigm of sustainable agriculture. European Congress on Conservative Agriculture, Madrid, 1-46.

Khan, N., Ray, R. L., Sargani, G. R., Ihtisham, M., Khayyam, M. and Ismail, S. (2021). Current progress and future prospects of agriculture technology: Gateway to sustainable agriculture. *Sustainability* 13(9): 1–31.

Korthals, M. (2016). *Encyclopedia of global bioethics: Agricultural Ethics*. Dordrecht: Springer Science.

Korthals, M. (2017). Is intensive farming ethical acceptable? Agricultural research and technology. *Open Access Journal* 7(2): 33–42.

Kuhn, T. S. (1970). *The Structure of Scientific Revolutions*. Chicago: University of Chicago Press.

Kuiken, T., Barrangou, R. and Grieger, K. (2021). Broken promises of sustainable food and agriculture through new technologies: The CRISPR Case. *The CRISPR Journal* 4(1): 25–31.

Kumar, R., Mishra, J. S., Mondal, S., Meena, R. S., Sundaram, P. K., Bhatt, B. P., Pan, R. S., Lal, R., Saurabh, K., Chandra, N., Samal, S. K., Hans, H. and Raman, R. K. (2021). Designing an ecofriendly and carbon-cum-energy efficient production system for the diverse agroecosystem of South Asia. *Energy* 214: 118860.

Lehretz, G. G., Sonnewald, S., Lugassi, N., Granot, D. and Sonnewald, U. (2020). Future proofing potato for drought and heat tolerance by overexpression of hexokinase and SP6A. *Frontiers in Plant Science* 11: 614534.

Lewis, J. (2020). Experimental design, ethics, integrity, and the scientific method. In: Ipohfen, R. (Ed.), *Handbook of Research Ethics and Scientific Integrity*. Switzerland: Springer Nature.

Livingstone, G. M. (2016). Community in Argentina fights fracking site. *BBC NEWS*. Available at: http://www.bbc.com/news/world-latin-america-36892770.

Luca, A. I. D., Molari, G., Seddaiu, G., Toscano, A., Bombino, G., Ledda, L., Milani, M. and Vittuair, M. (2015). Multidisciplinary and innovative methodologies for sustainable management in agricultural systems. *Environmental Engineering and Management Journal* 14(7): 1571-1581.

Mark, R. (2019). Ethics of using AI and big data in agriculture: The case of a large agriculture multinational. *ORBIT Journal* 2(2): 1–27.

Martos, V., Ahmad, A., Cartujo, P. and Ordonez, J. (2021). Ensuring agricultural sustainability through remote sensing in the era of agriculture 5.0. *Applied Sciences* 11(13): 1–26.

Mollah, T. H., Tahsin, M., Mollah, N., Mohammad, N., Hasan, M. R. and Nihei, T. (2021). Exploring role of local NGO's in agriculture: A GIS based study along the Jamuna River, Bangladesh. The Jahangirnagar review: Part II. *Social Science* XLV: 151-170.

Monshipouri, M., Welch, Jr., C. E. and Kennedy, E. T. (2003). Multinational corporations and the ethics of global responsibility: Problems and possibilities. *Human Rights Quarterly* 25(4): 965–989.

Özdağ, U. (2011). Sessiz Bahar'dan sonra ses getiren elli yıl: kadın, çevre, sağlık (Smashing fifty years after *the silent spring*: woman, environment, health). *Hacettepe Üniversitesi Edebiyat Fakültesi Dergisi* 28(2), s. 179–199 (in Turkish).

Passini, E., Britton, O. J., Lu, H. R., Rohrbacher, J., Hermans, A. N., Gallacher, D. J., Greig, R. J. H., Bueno-Orovio, A. and Rodrigez, B. (2018). In silico drug trials demonstrate higher accuracy than animal models in predicting clinical pro-arrhythmic cardiotoxicity. *Frontiers in Physiology* 8(688): 1–15.

Porter, E. M. and Kramer, M. M. (2007). Corporate social responsibility: Wohltaten mit system. *Harvard Business Manager, January 2007*, s. 16-34.

Pusztai, A. (1998). *Eat up your genes*. World in Action TV Series, Granada Television, Producer Julia Berg, Programme Editor Jeff Anderson, Great Britain. Available at: http://ftvdb.bfi.org.uk/sift/title/596535 (accessed: 22 August 2021).

Pusztai, A. (2003). Hot Potato 79. In: Rowell, A. (Ed.), *Don't Worry [It's Safe to Eat]*. London: Routledge.

Qui, D., Hu, W., Zhou, Y., Xiao, J., Hu, R., Wei, Q., Zhang, Y., Feng, J., Sun, F., Sun, J., Yang, G. and He, G. (2021). TaASR1-D confers abiotic stress resistance by affecting ROS accumulation and ABA signaling in transgenic wheat. *Plant Biotechnology Journal* 19(8): 1588-1601.

Ramesh, P. (2008). *Ethics in Agricultural Research*. 105th FoCARS Foundation Course for Agricultural Research Service.

Richardson, M., Coe, R., Descheemaeker, K., Haussmann, B., Wellard, K., Moore, M., Maland Cady, J., Gubbels, P., Tchuwa, F., Raz, R. and Nelson, R. Y. (2022). Farmer research networks in principle and practice. *International Journal of Agricultural Sustainability* 20(3): 247-264.

Ryan, M. (2020). Agricultural big data analytics and the ethics of power. *Journal of Agricultural and Environmental Ethics* 33(1): 49-69.

Sahil, S., Keshan, R. and Patra, A. (2011). Expression and regulation of stress-responsive genes in plants under harsh environmental conditions. In: *Harsh Environment and Plant Resilience*. Springer.

Sanders, K. T. and Webber, M. E. (2014). A comparative analysis of the greenhouse gas emissions intensity of wheat and beef in the Unites States. *Environmetal Research Letters* 9(4): 1-9.

Şanlı, B. A. and Gökçe, Z. N. Ö. (2021). Investigating effect of miR160 through overexpression in potato cultivars under single or combination of heat and drought stresses. *Plant Biotechnology Reports* 15: 225-348.

Shabbir, R., Javed, T., Afzal, I., Sabagah, A. E., Ali, A., Vincente, O. and Chen, P. (2021). Modern biotechnologies: Innovative and sustainable approaches for the improvement of sugarcane tolerance to environmental stresses. *Agronomy* 11(6): 1-20.

Shahare, M. and Roberts, L. L. (2020). Historicizing the crisis of scientific misconduct in Indian science. *History of Science* 58(4): 485-506.

Simonin, D. and Gavinelli, A. (2019). The European Union legislation an animal welfare: State of play, enforcement and future activities. In: Hild and Schweitzer, L. (Eds), *Animal Welfare: From Science to Law*, 59-70.

Singer, P. (1990). *Animal Liberation*. New York, NY: New York Review of Books.

Sztybel, D. (2006). Can treatment of animals be compared to the holocaust? *Ethics and the Environment* 11(1): 97-132.

Thompson, P. B. (1988). Ethics in agricultural research. *Journal of Agricultural Ethics* 1(1): 11-20.

Thompson, P. B. and Kaplan, D. (2014). *Agriculture Ethics, Encyclopedia of Food and Agriculture Ethics*. Dordrecht: Springer.

Thompson, P. B. and Stout, B. A. (2018). *Beyond the Large Farm: Ethics and Research Goals for Agriculture*. New York: Routledge.

Trnka, M., Hayes, M., Jurecka, F., Bartossova, L., Anderson, M., Brazdil, R., Brown, J., Camarero, J. J., Cudlin, P., Dobrovolny, P., Eitzinger, J., Feng, S., Finnessey, T., Gregoric,

G., Havlik, P., Hain, C., Holman, I., Johnson, D., Kersebaum, K. C., Ljungqvist, F. C., Luterbacher, J., Micale, F., Hartl-Meier, C., Mozny, M., Nejedlik, P., Olesen, J. E., Ruiz-Ramos, M., Rötter, R. P., Senay, G., Vicente-Serrano, S. M., Svoboda, M., Susnik, A., Tadesse, T., Vizina, A., Wardlow, B., Zalud, Z. and Büntgen, U. (2018). Priority question in multidisciplinary drought research. *Climate Research* 75(3): 241–260.

Truong, H. A., Lee, S., Trinh, C. S., Lee, W. J., Chung, E. H., Hong, S. W. and Lee, H. (2021). Overexpression of the HDA15 gene confers resistance to salt stress by the induction of NCED3, an ABA biosynthesis enzyme. *Frontiers in Plant Science* 12: 640443.

Turnbull, C., Lillemo, M. and Hvoslef-Eide, T. A. K. (2021). Global regulation of genetically modified crops amid gene edited crop boom- review. *Frontiers in Plant Science* 12: 1–19.

Tuvel, R. (2015). Against the use of knowledge gained from animal experimentation. *Societies* 5(1): 220–244.

Wegren, S. K. (2021). Prospects for sustainable agriculture in Russia. *European Countryside* 13(1): 193–207.

Welch, R. M. and Graham, R. D. (1999). A new paradigm for world agriculture: Meeting human needs. *Field Crops Research* 60(1–2): 1–10.

Wezel, A., Casagrande, M., Celette, F., Vian, J.-F., Ferrer, A. and Peigné, J. (2014). Agroecological practices for sustainable agriculture: A review. *Agronomy for Sustainable Development* 34(1): 1–20.

World Watch Institute (WWI) (2011). *Innovations that Nourish the Planet*. Available at: http://blogs.worldwatch.org/nourishingtheplanet/research/state-of-the-world-2011-2/ (accessed: 22 August 2021).

Yang, X., Lu, M., Wang, Y., Wang, Y., Liu, Z. and Chen, S. (2021). Response mechanism of plants to drought stress. *Horticulturae* 7(3): 1–36.

Yu, Y. and Zhang, L. (2021). Overexpression of TaWRKY46 enhances drought tolerance in transgenic wheat. *Cereal Research Communications* (2011).

Zampati, F. (2019). Providing an ethical approach to big data in agriculture. Available at: https://spore.cta.int/en/opinions/article/codes-of-conduct-providing-an-ethical-approach-to-big-data-in-agriculture-sid0700a1b64-6f93-4c70-aa77-186c152b2b1d (accessed: 22 August 2021).

Zietman, A. L. (2013). Falsification, fabrication, and plagiarism: The unholy trinity of scientific writing. *International Journal of Radiation Oncology, Biology, Physics* 87(2): 225–227.

Zimdahl, R. L. (2012). *Agriculture's Ethical Horizon*. London: Elsevier Publication.

Index

Ingram Content Group UK Ltd.
Milton Keynes UK
UKHW020126190623
423621UK00001B/4

9 781801 463133